Hi-tech Horticulture

Hi-tech Horticulture

K.V. Malam

Agromet cell, Department of Agronomy,
College of Agriculture, Junagadh Agricultural University,
Junagadh - 362001, Gujarat

V.R. Malam

Agromet cell, Department of Agronomy,
College of Agriculture, Junagadh Agricultural University,
Junagadh - 362001, Gujarat

D.R. Kanzaria

Department of Horticulture, College of Agriculture,
Junagadh Agricultural University, Junagadh - 362001, Gujarat

CRC Press
Taylor & Francis Group
Boca Raton London New York

CRC Press is an imprint of the
Taylor & Francis Group, an **informa** business

-EPH-

First edition published 2024
by CRC Press
4 Park Square, Milton Park, Abingdon, Oxon, OX14 4RN

and by CRC Press
2385 NW Executive Center Drive, Suite 320, Boca Raton FL 33431

British Library Cataloguing-in-Publication Data
A catalogue record for this book is available from the British Library

Print edition not for sale in India

ISBN: 9781032690599 (hbk)
ISBN: 9781032690605 (pbk)
ISBN: 9781032690612 (ebk)

DOI: 10.4324/9781032690612

Typeset in Adobe Caslon Pro
by Elite Publishing House, Delhi

-EPH-

Contents

Green manure
    Methods of Green Manure
    Characteristics desirable in legume green manure crops
    Advantage of Green Manuring
    Limitations of green manure

No-cost inputs
    Indicator plants
    Use of the planting calendar
    Homa therapy or agnihotra

Preface

Hitech horticulture is a new subject of study in the field of horticulture and a chain system of cultivation having proper linkage right from selection of seed variety for sowing to supplying it to the end user. Intensive cultivation, less environment dependent with hitech production input is the necessity for self-sufficiency in food. However, horticultural industry has reached unprecedented growth in the size, diversity and value worldwide, but still horticultural science within academia is experiencing a crisis and no serious attention has been paid for popularizing the recent development in horticultural crops. So, this book is an admirable attempt in this direction. Horticulture needs to redefine and/or focus itself because horticulture is an industry and its aim is economic success. The book consist of 17 chapters including (1) Hi-tech Horticulture: An Introduction, (2) Greenhouse technology: Introduction, Types of greenhouses, (3) Planning and Design of Greenhouse, (4) Greenhouse Irrigation Systems (5) Passive Solar Greenhouse, (6) Hi-Tech Nursery Management, (7) Problems/ Constraints of Greenhouse Cultivation and Future Strategies, (8) Growing Media, Soil Pasteurization, Cost Estimation and Economic Analysis, (9) Hydroponics for Hi-Tech Horticulture, (10) Plant Growth Regulators for Hi-Tech Horticulture, (11) Plasticulture for Hi-Tech Horticulture, (12) Artificial Intelligence, IoT and Robotics for Hi-Tech Horticulture, (13) Integrated Nutrient Management for Hi-Tech Horticulture, (14) Tissue culture for Hi-tech Horticulture, (15) Production Technology under Hi-tech Horticulture, (16) Drying methods and applications for Hi-tech Horticulture and (17) Good Manufacturing Practices (GMP) for Hi-tech Horticulture.

Presentation of tables, figures attracts more to the readers and even in a quick look one can get vital information. Preface Due to rich diversity of agro-climatic conditions and socio-cultural conditions prevailing in the country, India is regarded as a horticultural paradise. There needs to be an expansion and/or introduction of new high-income careers in horticulture. We hope that the text of the present book will be helpful to horticulturists/growers of India to increase export potential. The authors are thankful to various sources from where the literature is cited. Suggestions for improvement of this text book will be highly appreciated. It is hoped that the book 'Hi-tech Horticulture' will prove of immense value not only to research workers but also to the teachers, students, planner, farmers and individuals who are desirous of increasing quality fruit production in horticultural crops. Besides it is also expected to be useful to the horticulturists all over the world, seeking latest information on horticultural technologies.

Authors

About the Authors

Dr. K.V. Malam, senior author of this book received M.Sc. (Agri.) degree in Agronomy in 2019 and Ph.D. (Agri.) Agronomy in 2022. He commenced his professional career in 2019 as Young Professional-I and senior research fellow. He is actively engaged in teaching of U.G. and research activities. He has published several research papers, popular articles and books to his credit. He has also presented his research papers in National seminars/conferences. His present interest is in U.G. teaching and research particularly in Agronomy, Agricultural Meteorology and Horticulture.

Dr. V.R. Malam, author of this book received M.Sc. (Agri.) and Ph.D. degrees in Horticulture. He has been engaged in research and post-graduate teaching since last thirty three years. He received the Sardar Patel Agriculture Research award, Bharat Ratna Mother Teresa award and Eminent scientist award. He has published several research papers, popular articles and two books to his credit. He has been life member of half a dozen scientific societies, and has guided many post-graduate students.

Dr. D. R. Kanzaria, author of this book received M.Sc. (Hort.) and Ph.D. degrees in Horticulture in 1999 and 2015, respectively. He is actively engaged in teaching of U.G. and P.G. programmes, research, extension activities and has guided many post-graduate students. He has to his credit number of research papers, popular articles and three books. He has also presented his research papers in many National and International seminars/conferences.

Chapter - 1

Hi-tech Horticulture: An Introduction

HORTICULTURE AND HI-TECH HORTICULTURE

The term 'agriculture' derives from the Latin terms 'agri' or 'ager' meaning 'soil' and 'cultura' meaning 'cultivation.' Agriculture has all aspects of plant cultivation, livestock, fisheries and forestry and is very broad term. Horticulture is the agricultural branch related to intensively cultivated plants directly used by humans for food, medicinal purposes or aesthetic pleasure (Janick, 1972).

Definitions

Horticulture is the applied science and that term was first used in the 17th century. Horticulture is derived from two Latin words 'hortus' meaning 'garden' or 'enclosure' and 'cultra' meaning 'cultivation.' Horticulture is known collectively as garden culture or garden crop culture. Garden as such is a broad term itself. Garden comes from the Latin term 'Gyrdan' which means 'to enclose.' The planting of flowers, fruit, vegetables in small plots utilizing intensive farming techniques, the most intense type of horticulture is perhaps crop planting. India has strong potential to develop a broad variety of horticultural crops due to large variation in location, temperature, soil and other agro-climatic conditions. Modern horticulture may be referred as a part of agricultural science, which deals with the production, utilization and improvement of fruits, vegetables, flowers, ornamentals, plantation crops, medicinal and aromatic plants etc.

Hi-tech horticulture may be referred as technology that is modern, less environmentally dependent, capital-intensive and capable of improving productivity and quality in order to get more money. Hi-tech horticulture is now commonly used for manufacturing horticultural goods on a sustainable sector.

History of Horticulture

Fruits are mentioned in Vedas, in Purans, in Upnishads. The history of the growing of fruit is as long as humanity. Pomegranate (3500 B.C), Grapes (2440 B.C), Banana, Coconut and Mangoes (2000 B.C), Peach and Almond (1300 B.C), Olive (100 B.C) are said to be the oldest fruit. In 1056 A.D., China published its first book on litchi cultivation.

Commercial horticulture has recent origin in India (100 years old). In ancient days, orchards were grown for mostly hobby by the Kings Akbar, who planted > 1 lakh fruit trees in Dharbanga (in Bihar) called Lakhbagh in a book called Ain-e-Akbari. Mughals established gardens in Mughal.

Now a day, on a commercial scale, fruits, vegetables and flowers etc. are growing large areas in open fields. Garden crops usually contain only fruits, herbs and flowers. Yet today's horticulture is not only concerned with fruits vegetables and flowers but also other essential crops such as planting crops medicinal and aromatic etc., using and improving these crops. Modern horticulture can therefore be defined as part of agricultural science, which deals with fruit, vegetables, flowers, ornamentals, spices and condiments, planting crops, medicinal and aromatic plants, and so on.

Origins of Horticultural Science

Some horticultural historians think Egypt was the birthplace of horticulture. The roots of horticultural science originate from three things coming together:

1. Formation of scientific societies during the 17th century.

2. Establishment of agricultural and horticultural societies during the 18th century, and

3. Establishment of state-supported agricultural work in different countries throughout the 19th century.

»   Two famous horticultural societies active in England:

   »   The London Horticultural Society (later the Royal Horticultural Society) was founded in 1804 and,

» In 1903, the Society for Horticultural Science (later the American Society for Horticultural Science) founded.

Middle ages: Scientific Horticulture

The separation of horticulture from agriculture as a separate practice is commonly dated in Europe from the middle ages. Although there are many similar practices in horticulture and agriculture (weeding, fertilizing, watering, etc.), horticulture is different from agriculture due, for example, to its advanced practices, grafting, and the smaller scale of operations.

Horticulture as a Science

Horticulture is an art, science, technology of plant cultivation for human use. It is performed in a garden from the individual level down to the commercial level. It includes plants for food crops, fruits, vegetables, flowers, mushrooms, medicinal herbs, food herbs and non-food crops, trees and shrubs, turf grass and hops in its activities. The diversification of food, medicine, climate, social goods and services are all important for the growth and protection of human health and well-being. A gardener is an individual who tends toward a gardening, and therefore is a gardener, but not all gardeners.

Branches of Horticulture

Divisions of horticulture

On the basis of the crops dealt with and also their function and usage, the horticulture field is listed for convenience in the following sections.

1. **Pomology:** Two Latin words '*pomum*' meaning fruit and '*logos*' meaning science. So, pomology is science related to cultivation of fruit crops. **E.g.:** Grape, Mango, Guava, Sapota, Banana etc.

2. **Olericulture:** The term 'Olericulture' originates from the Latin word 'oleris,' which means pot herbs (vegetables) and the English word 'cultra,' which means to grow. Olericulture, therefore, simply means production of potherb. It is commonly used today to mean vegetable cultivation. **E.g.:** Brinjal, Okra, Tomato, Pumpkin etc.

3. **Floriculture:** The term floriculture is derived from the Latin word '*florus*' meaning flower and '*cultra*' meaning cultivation. So, study of flower crops means floriculture.

There are two sub-divisions of floriculture again.

Commercial Floriculture: Commercial floriculture deals with the cultivation of commercially produced, for profit (income) flower crops. **E.g.:** Aster, Carnation, Jasmine, Rose and Marigold etc.

Ornamental Floriculture It deals with the production of flower crops for ornamental, pleasure and fashion. **E.g.:** Balsam, Cosmos, Dahlia, Gerbera, Hollyhock, Hibiscus, Nerium, Poinsettia and Gaillardia, Zinnia etc.

4. **Arboriculture:** Arboriculture is concerned with growing annual trees intended for shade, avenue or ornamental purposes. **E.g.:** Cassia, Gulmohar, Polyalthia, Spathodea, etc.

5. **Plantation crops:** Planting crops are grown on a wide scale in large contiguous areas, owned and operated by a person or a corporation and the commodity used is only after processing. **E.g.:** Tea, Coffee, Coconut, Rubber, Cocoa etc.

6. **Spices and condiments:** Cultivation of crops whose development is mainly used for seasoning and flavouring dishes.

   Spices: Spices are unique products of plants which are used to add fragrance and flavor to food adjuncts. **E.g.:** Pepper, Cardamom, Clove, Cinnamon, All spice etc.

   Condiments: Condiments are those plant products that are mainly used to add flavor to the food adjuncts. **Eg.** Turmeric, Ginger, Red chillies, Onion, Garlic etc.

   All spices and condiments contain essential oils which contain little nutritional value and provide aroma, flavour and taste.

7. **Medicinal and aromatic plants:** It deals with the cultivation of medicinal plants, which provide drugs and aromatic crops which yields aromatic (essential) oils.

   Medicinal plants- Plants are highly rich in secondary metabolites and are potential sources of drugs. The secondary metabolites include alkaloids, coumarins, flavonoids, glycosides and steroids etc. **E.g.:** Menthi, Sarpagandha, Aswagandha, Tulasi, Dioscorea Yam, Belladona, Senna, Opium, Periwinkle, Cinchona etc.

   Aromatic plants- Plants possesses essential oils in them, which oils are the odoriferous steam volatile constituents of aromatic plants. **E.g.:** Citronella, Lemon grass, Vetiver, Davanam, Lavendor, Palmrosa, Geranium etc.

8. **Postharvest technology:** It deals with the postharvest handling, grading, packaging, storage, processing, preservation, value addition and marketing of horticulture crops.

Hi-Tech Horticulture in India

Horticulture is not only a means of diversification in the recent era, it also forms a vital part of food and nutritional security and is a significant element of economic security as well. Horticultural adoption has brought development in many states like Maharashtra, Karnataka, Kerala and Andhra Pradesh.

Judging by the dramatic rise in population, the added pressure on natural resources is great with global warming and climate change, declining land holdings and high demand for quality fresh horticultural produce. Situation requires a shift to modern crop production technologies, in which hi-tech horticulture has already taken the lead.

Term hi-tech horticulture refers to the precise manufacturing techniques for efficient use of inputs at the appropriate time and quantity to maximize yield and quality in various horticultural crops. It is an adoption of any technology that is modern, less environmentally dependent, capital-intensive, and has the capacity to improve horticultural crop productivity and quality. It is a chain system of fruit, flowers, vegetables and spices growing with a proper relationship from seed/ variety selection to final product through the process of modern crop production technique through post-harvest management techniques. Hi-tech horticulture has overcome an agro-climate barrier which makes most of the vegetables and other horticultural products available to consumers at a premium price throughout the year.

Hi-tech horticulture innovations include the Integrated Nutrient Management (INM), Integrated Pest Management (IPM), plasticulture, greenhouse or protected cultivation, hydroponics, microirrigation or drip irrigation, fertigation etc.

Integrated Pest Management (IPM)

Integrated Pest Management (IPM) is now becoming a methodology widely used in hi-tech horticulture. Integrated Horticultural Pest Management is one of the main criteria for supporting productive cultivation and rural growth. Integrated Pest Management seeks to utilize environmental, biological and chemical protection of pests and pathogens in a judicious manner.

Integrated Nutrient Management (INM)

Integrated Nutrient Management (INM) has now also become a widely practiced practice of hi-tech horticulture. Integrated Nutrient Management (INM) relates to managing soil fertility and the availability of plant nutrients to an optimal degree to ensure the required crop production by maximizing the benefits from all available plant nutrient sources in an integrated way. Another essential feature of INM is to improve the productivity of fertilizer use efficiency (FUE) through properly positioning fertilizer near to the maximum root operation rhizosphere. Integrated Nutrient Management has now become one of the standard practices for progressive producers of horticultures.

Plasticulture

Today, plastic farming has become a common technology for hi-tech horticulture. Plastics have various uses in the industrial processing of horticultural products. The method of using plastics to produce commercial horticulture is called 'plastic manufacturing.' Protected cultivation (high and low tunnels, etc.); plastic mulching and plastic lining are various applications of plastics in horticulture. Plasticulture increases the economic performance of production processes and assists in the effective use of water and electricity. Plasticulture decreases variations in temperature and moisture, and also helps to avoid infestations of pests and diseases. Plasticulture plays a dominant role in the precise use of irrigation and nutrients by reducing water and nutrient waste and reducing soil erosion. The use of plastics has been shown to promote the sound use of natural resources such as soil, water, sunlight and temperature.

Greenhouse Cultivation

Greenhouse cultivation or protected cultivation is currently very popular among progressive horticultural growers. This technology of hi-tech horticulture provides many advantages over conventional production techniques such as in greenhouse cultivation, horticultural products can be grown mainly fruit, vegetables and flowers under protected cultivation even during off-seasons.

Advantages of Greenhouse are

» Production of vegetable crops.

» Production of roses, carnation, cut-flowers etc.

» Plant propagation, raising of seedlings.

» Primary and secondary hardening nursery of tissue cultured plant.

» Growth/production of rare plants, orchids / herbs, medicinal plants

Hydroponics

Hydroponics is another technology of hi-tech horticulture which offers great scope to horticultural producers worldwide. The soilless cultivation is also known as hydroponics. Hydroponics helps producers grow plants in a solution of nutrients, without using the normal soil media.

Drip or micro-irrigation irrigation

Drip irrigation is now a commonly used method in worldwide irrigation. There have been some advantages of drip irrigation over the conventional irrigation system. Such benefits include optimal usage of irrigation water, maximum productivity of water use by supplying water inside the plants' root system, and low evaporative loss of soil moisture.

Fertigation

The practice of supplying plant fertilizers and nutrients through irrigation water is known as fertilization. Fertigation is normally achieved with irrigation by the drip.

Horticultural Scope

For the following reasons, horticulture can be defined as having great scope:

1. To exploit the great unpredictability of the conditions of agro-climates.

2. To satisfy the rising population need for fruits, vegetables, flowers, spices and beverages.

3. To satisfy industry's need for processing.

4. To increase supplementary imports and exports.

5. To improve employment, farmers, entrepreneurs' economic conditions and to engage more labourers to solve the unemployment problem.

6. Horticultural development requires specialized skills such as manufacturing, harvesting, marketing, canning, refining and many other related trades; hence it can handle a huge workforce

7. To protect environment.

Chapter - 2

Greenhouse Technology

What is Protected Cultivation?

Protected cultivation can be referred as cropping technique in which the microclimate surrounding the plant body is partially or fully regulated as required by the plant species grown during their growth time.

Importance and Scope of Protected Cultivation

Protected cultivation is the best choice for quality produce and efficient use of land and other resources in the current scenario of continuous demand for better quality vegetables, fruits and flowers continually shrinking soil holdings. Growers don't get good returns during the on season due to the wide abundance of vegetables in the markets. For example, biotic stresses during the rainy and post rainy seasons do not permit effective vegetable production in several parts of the country. As a result, most vegetables are affected by a serious occurrence of diseases, pests and viruses, thereby affecting vegetable quality. Given the pressure of increasing population, rapid urbanization and industrialization, cultivable area is becoming limited day by day. Growing market for high-quality vegetables. Rising early in the nursery to get an early crop.

Principles of Protected Cultivation

The protected cultivation structure is a structure that is covered by a transparent material to allow natural light for plant growth. Greenhouse cultivation produces higher vegetable yield and other horticultural crops. In addition to rising growth, it also increases product consistency and ensures availability throughout the year

Protected growing technology is best suited for vegetable crops (cucumber, tomato, capsicum, chilies, etc.), flowers (rose, gerbera, carnation, etc.) and nursery for all vegetable crops due to their short lifespan. Greenhouse also delivers higher CO_2 concentrations to increase production to its maximum.

The following benefits can be obtained by monitoring the environmental parameters (i.e. temperature (°C), light intensity (lux) and relative humidity (percent)) and carbon conservation (CO_2) within the structure:

» Efficient soil, fertilizer and energy consumption.

» Under adverse climatic conditions, any crop can be grown.

» Reduce the cost of plant protection.

» Lower average crop yield (30-40 percent) than open field crops

» Enhancement of product quality.

» Rise in profits for farmers

Greenhouse Technology

Introduction

Use of greenhouse technology began in India during the 1980s and was primarily used for research activities, while greenhouses have existed in different parts of the world for more than one and a half centuries. Following the introduction of the green revolution, more importance is given to the quality of the commodity along with the quantity of output to meet the ever-increasing demands of food. All of these demands can be fulfilled when the plant growth condition is properly controlled. The need to protect the crops from unfavourable environmental conditions has led to the development of protected agriculture. Greenhouse is the most realistic way to achieve safe agriculture goals, where the natural environment is modified by sound engineering principles to achieve optimum plant growth and yields.

What is Greenhouse?

Plant growth is both an art and a science. Around 95 percent of plants are grown in the open area, either food crops or cash crops. A greenhouse is a broad term referring to the use of a transparent or partially transparent substrate that is protected by a frame to enclose an area for propagating or growing plants, whereas 'Glass house' refers to glass as the covering material. A 'greenhouse' or 'polyhouse' refers to the use of sheeting or plastic films. The structure is known as a 'shade house' when the enclosing material is woven or otherwise built to let sunlight, moisture and air pass through the gaps. The environmental conditions are so changed that any plant can be grown anywhere at any time by supplying minimal labour with suitable environmental conditions.

Greenhouse is an enclosed structure, covered by (transparent and translucent) glass or plastic film. Under this system, plants are cultivated under the partially or completely regulated climate. The greenhouse technology has been of tremendous importance for better use of energy, growing crops under adverse climatic conditions and high areas of rainfall. Since plastic film serves as selective radiation filters, the incoming short wave solar radiation passes through it and the outgoing long wave radiation partially traps the thermal energy in the greenhouse released by the objects contained in it, this phenomenon is known as the "Greenhouse Effect." Today, due to protected cultivation, it is possible to grow most seasonal crops throughout the year which allows us to control and manipulate the climate.

The greenhouse gases include carbon dioxide (CO_2), methane (CH_4), nitrous oxide (NO_2), chlorofluorocarbons (CFCs), hexafluoride to sulphur and others. It is estimated that 50% of the greenhouse effect is caused by CO_2, 20% by methane, 14% by CFCs and the remainder by other elements, including water vapours.

Greenhouse technology

For some temperate regions where the weather conditions are extremely adverse and no crops can be grown, man has created methods of continuously growing some high-value crop by providing protection from the intense cold called Greenhouse Technology. Greenhouse technology involves building design that can take care of climate control mechanisms to correct plant microclimate and also required for optimum plant growth and development in this house.

Greenhouse technology can be described as the technology required to create a building, partially covered, completely or porously to regulate the indoor climatic conditions under which crops / plants can be grown using special technology.

Greenhouses are framed or inflated buildings covered with transparent or translucent material that is sufficiently wide to grow crops under partial or fully controlled environmental conditions to achieve maximum growth and productivity.

Advantages of greenhouse Technology

» Gives the plants favourable micro-climate.

» Plus perfect for high-value and off-season cultivation.

» Increases yield per unit area, with better quality.

» Severely limited and easy to monitor crop water requirements.

» Good use of chemicals, insect control poisons and diseases;

» Helps to harden plants grown in tissue and to increase early nurseries;

» Dissemination of planting material is available year-round.

» Protects the crops against adverse conditions.

» Brings opportunities for self-employment to educated youth.

» Depending on the type of greenhouse, crop size, environmental control facilities, the yield can be 10-12 times greater than outdoor cultivation.

» Modern techniques of hydroponic (less soil culture), aeroponics and film techniques with nutrients are only possible under greenhouse cultivation.

Disadvantages of greenhouse technology

Production of protected plants in the country is very ancient. Very little effort was made to exploit the tremendous potential of safe development. This technology involves engineers, physiologists, physicists, horticulturists, environmentalists, etc. working together skilled people.

Problems requiring concerted efforts are:

» To different agro-climatic regions of the world, uniform greenhouses and other design structures are not appropriate.

» Resources for loading are very expensive. Low-density ultraviolet polyethylene is not easy to get. Intensive work is needed to make the cladding material acceptable for growers. Plastic films are expensive and have a short life in the world.

» Plant production technologies for different crops are not developed and recorded under different types of protected structures for different agro-climatic areas of the world.

» Absence of a primary safe farming research programme.

» Lack of awareness among farmers concerning safe output potentials.

Classification of Greenhouses

For crop production various types of greenhouse structure are used. There are advantages for a specific application in each type, but there is generally no single greenhouse type, which can be considered as the best. Numerous types of greenhouses are designed to fulfill the different needs. The various types of greenhouses classified in terms of shape, utility, construction and covering material are briefly described below:

A. Greenhouse type based on shape

Greenhouses can be categorized according to shape or design. Uniqueness of the greenhouse cross section can be considered as classification factor. Since the longitudinal section is almost similar for all forms, only the greenhouse longitudinal section cannot be used for classification. The cross-sections show the structure width and height, and the length is perpendicular to the cross-section axis. The shape-based types of greenhouse commonly followed are lean-to, even span, uneven span, ridge and furrow, saw tooth and Quonset.

1. Lean-to type greenhouse

At that time lean-to-type design is used when a greenhouse is placed against the side of an existing building. This is designed against a building and uses the current frame on one or more of its sides. It is normally attached to a house or may be attached to other buildings. The building's roof is extended with sufficient greenhouse cover material, and the area is fully enclosed. For sufficient sun exposure it usually faces south side. With a total width of 7 to 12 feet, this style of greenhouse is limited to single or double-row plant benches. This can be as long as the building to which it is connected.

This is usually near to buildings, so that electricity, water and heat are more available. The system is the least costly. Lean-to type greenhouse allows maximum use of sunlight and minimises the roof support requirements. These are the Lean-to-Greenhouse advantages. The height of the supporting wall restrains the greenhouse's potential capacity. Limited space, limited light, limited ventilation; and temperature control is very difficult because the wall on which the greenhouse is built may collect the heat from the sun while the greenhouse's translucent cover may lose heat rapidly. Those are the lean to style greenhouse drawbacks.

2. Even span type greenhouse

These kinds of greenhouse are standard type and full-size structure, because the pitch and width of the two roof slopes equals. This style is used for the greenhouse of limited scale, and it is designed on level land. This can accommodate 2 or 3 rows of benches with plants. An even-span greenhouse costs more than just the lean-to kind. The even-span would cost more to heat, due to its size and greater amount of exposed glass space. Also span style greenhouse has a better shape to maintain uniform temperatures during the winter season than lean-to model for air circulation. It will house two side benches, two walks and a broad bench in the middle. In several regions of India single and multiple span types are developed. In general, the span ranges from 5 to 9 m for the single span style greenhouse, the length is about 24 m and the height varies between 2.5 and 4.3 m.

3. Uneven span type greenhouse

Greenhouse style uneven range is built on hilly regions. The roofs are unequal in width, so this system can be modified to use the hill's side slopes. Nowadays, this type of greenhouse is seldom used, because it is not adaptable for automation.

4. Ridge and furrow type greenhouse

Greenhouse style ridge and furrow use two or more A-frame greenhouses linked to each other along eave range. The eave acts as a furrow or gutter for clearing rain and melting snow. The side wall between the greenhouses is removed which results in a single large interior structure. Its design reduces energy, decreases labour costs, increases personal management and decreases fuel consumption due to less exposed wall area from which heat flows. In the frame specifications of these greenhouses, the snow loads must be taken into account, as the snow cannot slide off the roofs. The greenhouse style of ridge and furrow is commonly used in northern European and Canadian countries, and is well suited to Indian conditions.

5. Saw tooth type greenhouse

Apart from allowing for natural ventilation, saw tooth type greenhouse is similar to ridge and furrow style greenhouses. In this type of greenhouse develops specific natural ventilation flow pathway.

6. Quonset greenhouse

Quonset greenhouse, in which pipe arches or trusses are supported by pipe purlins running along the greenhouse length. For this type of greenhouses polyethylene is commonly used for the covering material. Generally these greenhouses are less costly than the greenhouses connected to the gutter and are ideal when requiring a small isolated cultural environment. These houses are either linked in a standing-up style or arranged in an interlocking ridge and furrow. In this type, there is a single large cultural space for a set of houses, that type of arrangement is better adapted to the labourers' automation and movement.

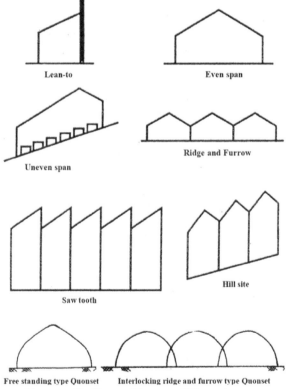

Fig. 2.1: Different Shapes of Greenhouses

B. Greenhouse type based on utility

Greenhouses on the functions or utilities can be classified. Utilities include greenhouse heating and artificial cooling. The greenhouses are also graded as greenhouses for the active heating system and the active cooling system based on that.

1. Greenhouses for active heating

During winter, indoor greenhouse air temperature decreases during the night time. Some amount of heat needs to be given to avoid the cold bite from freezing to plants. The greenhouse heating needs depend on how much heat is lost to the outside environment. A variety of techniques are used to minimize heat losses, i.e. by using double-layer polyethylene, thermo-pane glasses (two layers of factory sealed glass with dead air space) or by using heating systems such as solar heating systems, unit heaters, central heat and radiant heat.

2. Greenhouses for active cooling

In summer, for successful crop production, the temperature of the greenhouse is needed to be lowered than the ambient temperature. But in the greenhouse necessary modifications are made such that a large amount of cooled air is drawn into the greenhouse. This type of greenhouse consists of an evaporative cooling pad with cooling by fan or fog. This type of greenhouse is designed to allow 40 percent and in some cases nearly 100 percent opening of the roof.

C. Greenhouse type based on construction

The structural material is mainly influenced by this method of construction, though the covering material also influences the form. In addition, period of the house determines the selection and design of the structural members. As the span is higher the material should be heavier and more structural elements should be used to build solid truss style frames. Simpler designs such as hoops can be followed for smaller spans. Therefore, depending on the design, greenhouses may be divided into wooden framed, pipe framed and truss framed structures.

1. Wooden framed structures

In general only wooden framed structures with a span of less than 6 m are used for this greenhouse. Side posts and columns are made of wood without the use of a truss. Pine wood is usually used because it is less costly and has the necessary energy. Because of the strong strength, durability and machinability, timber wood can be used for construction where it is locally available.

2. Pipe framed structures

Galvanized mild steel pipes are used for construction of this type of greenhouse, and the transparent span is around 12 m. The side posts, columns, cross ties and purlins are designed using tubing, and no trusses are used either. The elements of the pipe are not interconnected, depending on the fitting to the support sash bars.

3. Truss framed structures

When the span is more than or equal to 15 m at that time the truss frames are used in the greenhouse. Flat steel, tubular steel or angular iron is welded together, forming a truss that includes rafters, chords, and struts. Struts are compressed support members and Chords are tensioned support members. At each truss are bolted angle iron purlins running throughout the length of the greenhouse. Columns are used at that time when the very large truss frame houses of 21.3 m or more are to be constructed. Mostly glass houses are truss frame type, as these frames are best suited for pre-fabrication.

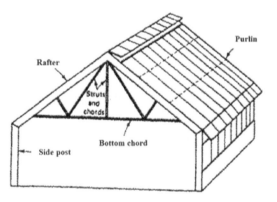

Fig. 2.2: Pipe Framed Greenhouse Structure

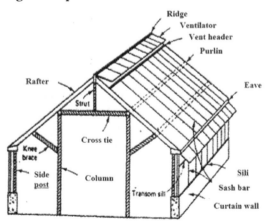

Fig. 2.3: Truss Framed Greenhouse Structure
Source: *http://www.eagri.org/eagri50/AENG252/lec02.pdf*

D. Greenhouse type based on cladding materials

Greenhouse is covered by cladding materials which are a major and important component of the greenhouse structure. The temperature of the air inside the house depends on the covering of the materials. The types of frames and the method of fixing are also different from the covering material. Depending on the type of cladding material, the greenhouses are classified as glass, plastic film and rigid panel greenhouses.

1. Glass greenhouses

Greenhouses with glass covering material existed before 1950. Glass as a covering material provides higher indoor light intensity, higher air infiltration rate, which leads to lower indoor humidity and better disease prevention. Lean-to type, even span, ridge and furrow type of designs are used for construction of glass greenhouse. This type of greenhouse is rarely used today because it is not adaptable for automation.

2. Plastic film greenhouses

Flexible plastic film, including polyethylene, polyester and polyvinyl chloride, is used as a covering material for this type of greenhouse. Plastics as a material for greenhouses have become very popular, because they are cheap and the cost of heating is lower than the cost of glass greenhouses. But plastic films are short-lived. For example, the best quality ultraviolet (UV) stabilized film can only be used for four years. The design of the Quonset as well as the gutter-connected design is suitable for the use of this covering material.

3. Rigid panel greenhouses

Rigid panels made of polyvinyl chloride (PVC), fibre-glass reinforced plastic (FRP), acrylic and polycarbonate rigid panels are used as covers. These can be used in frames of the Quonset or ridge and furrow type. This material is stronger to break and the light intensity is even higher in the entire greenhouse compared to glass or plastic. High-grade panels last for up to 20 years. However, the main drawback is that these panels tend to collect dust as well as harbor algae, resulting in the darkening of the panels and the resulting reduction in light transmission. There is a major risk of fire hazard.

E. Shading nets

That cultivated plant has to be provided the particular type of shade required for the various stages of its growth. The shading nets give the plants appropriate micro-climate conditions. Shade nets are meant to shield crops and plants from UV radiation, but they also provide protection against climatic conditions such as temperature variability, heavy rain and winds. Better growing conditions for the crop can be achieved because of the controlled micro-climate conditions "created" by shade netting in the covered area, which increases crop yields. All shade nettings in the exposure region are UV optimized to meet planned lifetime. They are divided into the break, low weight for simple, fast mounting and a shade range of 30 to 90%. There is a wide variety of shading nets on the market, which are dependent on the percentage of shade they provide to the plant developing under networks.

F. Classification of greenhouse based on suitability and cost

1. Low cost or low tech greenhouse

Low cost structure for the greenhouse with local resources, such as bamboo, wood etc. As a shielding medium, the UV film is used. There is no specific control instrument available in the greenhouse to regulate environmental parameters. Simple techniques for temperature and humidity increase or decrease are employed. When adding filtering materials such as nets, light intensity may also be popular. The temperature can be decreased during the summer by opening the side walls. The rain shelter for plant production is this kind of structure. When all the sidewalls are lined with plastic film, inside temperature is otherwise increased. This type of greenhouse is suitable primarily for cold weather zones.

2. Medium-tech greenhouse

This type of greenhouse is constructed using galvanized iron pipes. Greenhouse users prefer to have minimum investment in manual or semiautomatic control arrangements. With the support of the screws the canopy cover is attached with structure. To withstand the disturbance against wind, the whole system is firmly fixed with the ground. Temperature control is provided by exhaust fans with a thermostat. Also, evaporative cooling pads and misting arrangements are made to maintain in the greenhouse a favourable humidity. Since these devices are semi-automatic, they take a lot of care and attention, and maintaining a stable environment during the growing season is very challenging and bulky. These greenhouses are ideal for composite and dry areas of the climate.

3. Hi-tech greenhouse

The greenhouses of this type are fully automatic. The hi-tech greenhouse will solve difficulties with medium-tech greenhouse applications. The greenhouse can be filled with screening, heating, cooling and lighting equipment to maximize plant growth by means of a device.

Table 2.1: Types of greenhouse and their costs involved

Sr. No	Types of greenhouses	Cost/ SQ.mt. (Rs.)
1	Low-tech greenhouses (without fan and pad)	300-500
2	Medium-tech greenhouses (with fan and pad system without automization)	800-1100
3	Hi-tech greenhouses (with fully automatic system)	2000-3500

Chapter - 3

Planning and Design of Greenhouse

You need to carefully choose a greenhouse that best suits your needs. Some of the most critical moves a grower would need to take is to prepare for a new greenhouse. For the greenhouse and location in relationship to customers, labour, services and future growth, the right selection of the site is of extreme importance to affect the effectiveness of the business.

SITE SELECTION AND ORIENTATION

The right place is just as critical to choose for your greenhouse as your building materials, seeds, plant nutrients and various other components to ensure success. Because several things must be taken into consideration when choosing a site, ten considerations must be taken into consideration before making this crucial decision.

1. **Capability of expansion:** It is recommended to buy more land than you first wanted, so you have the chance to grow later.

2. **Elevation:** Elevation affects high summer and low winter temperatures, which can also affect the cost of cooling and heating. Tomatoes, for example, work best between 59 °F to 86 °F. Here are few tips on how to regulate the temperature if you happen to be at a higher altitude:

   » Place plastic water jugs around the plants when the temperature is high. During the day, the water will gather heat, and hold it in the night when the temperature drops.

» Place a greenhouse temperature alarm in the colder months to warn you when the temperature drops. If required, a ceramic heater can be installed inside the house to increase the temperature.

» Use an automatic vent opener to control heat if you have a rigid framework, and if you have a soft greenhouse framework, you can open a flap to allow heat reduction.

» Using an automatic water device during the summer, probably one with misters in it. Disconnect all hoses during the winter to avoid freezing of pipes, and water now by hand.

3. **Level and stable ground:** Be careful of the stability of the land you are building your greenhouse on; it should not be subject to moving. The field must also be graded for drainage of water (a drop of 6 inches per 100 feet). In addition, the ground must be compacted so it won't start sinking after the greenhouse is installed.

4. **Microclimate:** Your climate can be influenced by several different factors like latitude. At the equator, sea level at the poles will still be cooler than sea level, and large bodies of water will heat up and cool down much slower than land masses. For example, San Diego, which lies next to the Pacific Ocean, has much smaller fluctuations between day and night temperatures than the Sonoran desert, which is not close to a large body of water and whose temperatures may fluctuate.

Take trees, mountains and other obstructions that, particularly in the morning, might possibly cast a shadow on the greenhouse into account. Mountains are also capable of affecting wind and storm patterns. Other environmental factors include: clouds and fog that can collect in different areas at certain times of the day and minimize sunshine and photosynthesis; high wind that can cause structural damage and draw heat away from the greenhouse; swirling dust and sand that can braze greenhouse glazing and, last but not least, snow.

5. **Availability of labour:** The project owner or manager would require two distinct forms of labour force:

» Trainees as a retainable workforce. These laborers will take care of the harvesting and packaging of plants and fruits.

» Trained laborers. This includes farmers, managers of plant growth, plant nutrition specialists, plant safety specialists, office/computer specialists, specialists in labor/management and marketing specialists.

6. **Roads:** Clear road access is a must. For example, if roads are unpaved, the fruit will be subjected to all of the vehicle's rocky movement when you transport your harvest, which could result in bruising, crushing and other major damage to your fruit.

7. **Water:** Approximately 1 gallon of nutrient-mixed water is needed daily to supply each plant in addition to the water required for evaporative cooling, which is about 10,000 to 15,000 gallons per acre each day. To maximize the effectiveness of your water usage, you should recycle nutrient water, but be careful of salt build-up. To determine the salt and pH levels, an initial water analysis should be carried out. The pH levels for tomatoes should be changed to about 5.8 to 6.5 and, if the source water is essential, add acids such as nitric, phosphoric and sulfuric as in more than 7 parts per million. If the source water is acidic, then add a base as in less than 7.

8. **Utilities:** Make sure you have available the following utilities:

   » Telecommunication infrastructure

   » Three-phase electricity

   » Heating/ CO_2 generation combustible: natural gas, propane, fuel oil, electricity. Alternatives are wind, compost, woodchips, nut hulls and so on.

9. **Pest pressure:** Make sure you either select a site that is far from other areas of agricultural production, or create a buffer zone between your activity and other areas of production to avoid pest infestation.

10. **Orientation:** Correct orientation inside the greenhouse will offer good environmental conditions. Following points should be considered when deciding on a greenhouse's orientation, depending on light intensity and wind direction & velocity.

    » The single-span greenhouse orientation should be oriented towards East-West and North-South for multi-span use of the available sunlight.

    » Gutter should be produced in a multi-span greenhouse in the north-south direction.

» The slope along the gutter should not exceed 2%.

» The slope on the gable side does not exceed 1.25%.

» Ventilators should be opened on the leeward side in naturally ventilated greenhouse.

» The long axis of the single span greenhouse should be perpendicular to the wind direction to protect it from damage to the wind.

» Wind breaks should be placed on the north-west side of the greenhouse at least 30 meters away.

» In tropical regions, solar radiation is less available and sky remain cloudy during the rainy season. Consequently, effective ventilation is very important for managing temperature and humidity by maintaining the proper polyhouse orientation.

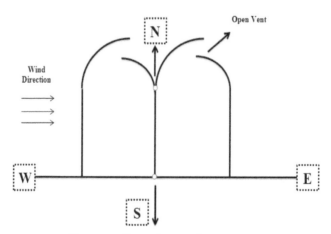

Fig. 3.1: Greenhouse Orientation

Structural Design of Greenhouse

The greenhouse structure's most significant purpose and its covering is to protect the crop against hostile weather conditions (low and high temperatures, snow, frost, rain and wind), diseases and pests. The construction of greenhouses with maximum intensity of natural light inside is important. The structural pieces in the greenhouse that can cast shadows should be minimised. The various greenhouse structural designs are available, based on frame styles. The most common design for a greenhouse may be a straight side wall with an arched roof but the gable roof is often commonly used. All structures may be either free standing or canopy attached to the greenhouse arch roof. Greenhouses in the arch roof and hoop style are most commonly built of galvanized iron pipe. If tall growing crops are to be grown in a greenhouse or when benches are used, it is best to use a straight side wall structure instead of a hoop style house which ensures the greenhouse's best operational use.

A greenhouse type hoop is suitable for low-growing crops such as lettuce, or for nursery stocks that are housed in greenhouses in extremely cold regions during the winter. A gothicarch frame structure can be built to provide the structure with sufficient sidewall height without loss of power. Heating and cooling devices, and water pipes. The greenhouse systems should be designed to withstand a wind speed of 130 km/h. The actual load depends on the wind angle, the shape and size of the greenhouse and the presence or absence of openings and breaks in the wind.

A greenhouse ultimately architecture depends on the following aspects:

1. The structural design as a whole and the properties of the individual structural components.

2. The basic mechanical and physical properties that decide the structural nature of the materials being covered.

3. The crop's unique exposure to light and temperature to be cultivated in the greenhouse.

4. The basic criteria which are applicable to the covering material's physical properties.

5. The Crop's agronomic requirements.

Structural components of Greenhouse

» **Side posts:** Sustain the trusses and support the greenhouse, usually 10 feet apart, in concrete footings.

» **Curtain wall:** The first few feet above the ground line on a sidewall. Usually made of solid building materials like concrete poured, concrete blocks, bricks, or lumber treated.

» **Sill:** The top of the curtain wall

» **Eave:** Where greenhouse sides are connected to the greenhouse wall. The top of the greenhouse's sides.

» **Truss:** Structural component supporting the roof's weight. The rafters, stretchers, and chords are included.

» **Purlin:** Purlin run the length of the greenhouse. Keep the roof trusses aligned.

» **Ridge:** Where the greenhouse's roofs are joined together. A lot of greenhouses have the winds of a ridge.

» **Sash bars:** Run perpendicularly to the purlins and attached to them holds the vitreous. Often built with a drip groove or a condensation channel that is formed within the glass panels.

Greenhouse Equipments

» **Roof:** transparent cover of a green house.

» **Gable:** transparent wall of a green house

» **Cladding material:** transparent material mounted on the walls and roof of a green house.

» **Rigid cladding material:** cladding material with such a degree of rigidity that any deformation of the structure may result in damage to it. Ex. Glass

» **Flexible cladding material:** cladding material with such a degree of flexibility that any deformation of the structure will not result in damage to it. Ex. Plastic film

» **Gutter:** collects and drains rain water and snow which is place at an elevated level between two spans.

» **Column:** vertical structure member carrying the green house structure

» **Purlin:** a member who connects cladding supporting bars to the columns

» **Ridge:** highest horizontal section in top of the roof

» **Girder:** horizontal structure member, connecting columns on gutter height

» **Bracings:** To support the structure against wind

» **Arches:** Member supporting covering materials

» **Foundation pipe:** Connection between the structure and ground

Span width: Centre to centre distance of the gutters in Multiplan houses

Green house length: dimension of the green house in the direction of gable

Green house width: dimension of the green house in the direction of the gutter

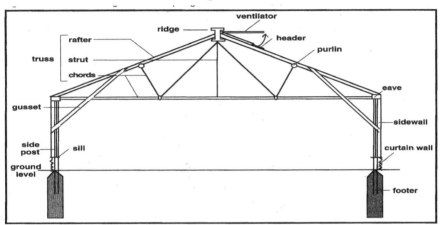

Fig. 3.2: Components of Greenhouse

Source: http://agritech.tnau.ac.in/horticulture/horti_Greenhouse%20cultivation.html

Design consideration for greenhouse:

The structure has to carry the following loads and is to be designed accordingly.

a. **Dead load:** Weight of all permanent building, insulation, heating and cleaning equipment, water pipes and all fixed frame service equipment.

b. **Live load:** Use overlapping weights (including hanging boxes, tables, and roof-workers). The greenhouse should be built with a live load not exceeding 15 kg per square meter. When applied at its middle, every roof member should be able to hold 45 kg of concentrated load.

c. **Wind load:** The winds of 110 km/hr and wind pressure of at least 50 kg/m² will be resistant to the structure.

d. **Snow load:** These will be taken according to the location's normal snowfall. A dead load, live load or dead load plus wind load, plus half live load should be taken by greenhouse.

The greenhouses are constructed from the Iron Pipes galvanized. Foundation in PCC with diameter 60 cm x 60 cm or 30 cm and depth 1 meter can be 1:4:8. PCCs with a thickness of 5 cm would also protect the vertical poles at a height of 60 cm. It stops the pole rusting.

Size: Based on the available ground, the greenhouse must be chosen. Depending on greenhouse types and number of suppliers in the area, the costs may vary. It is suggested to start with a naturally ventilated greenhouses with a minimum size of 100 m², depending on their market access and their experience with greenhouse cultivation, as less capital investment is necessary along with operating expenses. Depending on their extent of operations and project costs, however, experienced farmers / entrepreneurs can decide to take up a larger greenhouse.

Height: Highness is one of greenhouse design's main aspects. The structure's height has a direct influence on natural airflow, internal climate stability and crop management. For large greenhouses, the optimal centre height (up to 250 m²) for natural ventilated small greenhouses is 3.5 m to 4.5 m and 5.5 m to 6.5 m. In small and large greenhouses the side / gutter height will be between 2.5 m and 3 m and 4.5 m to 5 m. In single or multi-range structures both types of greenhouse can be made. For an area of over 200 m² a multi-speed greenhouse can be installed and is economical with regard to construction material and control / monitoring equipment required. The greenhouse's height should be marginally lower than the naturally ventilated greenhouse, and should not exceed 5.5 m in any situation.

Table 3.1: Type of structure

Component	Type of structure		
	NVPH Small (up to 250 m²)	NVPH large (> 250 m²)	Greenhouse with Fan & Pad system
Central height	3.5-4.5 m	5.5-6.5 m	<5.5 m
Side/ gutter height	2.5-3.0 m	4.5-5.0 m	As per NVPH size

**NVPH- Naturally Ventilated Polyhouse

Cladding materials: There are two types of plastic films normally used for greenhouse cladding - single layer (monolayer) clear transparent UV stabilized film and special films such as diffused, anti-drip, anti-dust, anti- sulphur etc. Different chemical processes performed during the cycle of production would depend on the crops produced in the greenhouse. The plastic films must have the following important properties according to the Indian Standard (IS 15827: 2009):

Table 3.2: Important properties of covering materials:

Sr. No.	Type of plastic films	Characteristics	Uses
i)	Normal film	a) Good Transparency more than 80%	Forcing and Semi- Forcing crops
		b) Low Greenhouse Effect	
ii)	Thermic clear film	a) Good Transparency	As normal film, when greater infrared (IR) effectiveness is desired.
		b) High infrared (IR) Effectiveness	
iii)	Thermic diffusion	a) Diffusion Light	As normal film, when greater infrared (IR) effectiveness and Diffusing light is desirable.
		b) High infrared (IR) Effectiveness	

Table 3.3: Comparison of different kinds of covering materials:

Type	Durability (year)	Transmission		Maintenance
		Light (%)	Heat (%)	
Polyethylene	01	90	70	Very high
Polyethylene UV resistant	02	90	70	High
Fiber glass	07	90	05	Low
Tedlar coated Fiber Glass	15	90	05	Low
Double strength glass	50	90	05	Low
Poly carbonate	50	90	05	Very Low

The ideal selective greenhouse material should have the following characteristics:

» The visible light portion of the solar radiation used by plants for photosynthesis should be transmitted.

» It should absorb little UV in the sunlight and convert some of it into visible light, which is good for plants, into fluoresce.

» The IR radiation that isn't beneficial to plants and causes the greenhouse interior to overheat will reflect or absorb.

» Will be ten to twenty years of age.

Design Criteria for Greenhouse Cooling and Heating Purposes

Greenhouse cooling

Greenhouse ventilation and greenhouse cooling are important in summer and winter. During summer, cooling by evaporative cooling systems and in winter cooling uses convection tube with pressurizing fans and exhaust fans. Cooling the greenhouse requires that the greenhouse has large air volumes.

a. Active summer cooling systems

Active summer cooling is accomplished by evaporative cooling. The two currently in use active summer refresher systems are fan-and-pad and fog systems. The cooling can only be done up to the wet bulb temperature of the incoming air in the evaporative cooling process.

1. Fan and pad cooling system

After 1954, the fan and pad evaporative cooling system has been available and is still the most popular summer cooling device in greenhouses. Water is passed through a pad along one greenhouse wall, which is normally mounted vertically in the wall. At the opposite wall there are the exhaust fans. Warm air is pulled in through the pad outside. By the evaporation cycle, the supplied water in the pad absorbs heat from the air that passes through the pad as well as from the pad and frame surroundings, thus triggering the cooling effect.

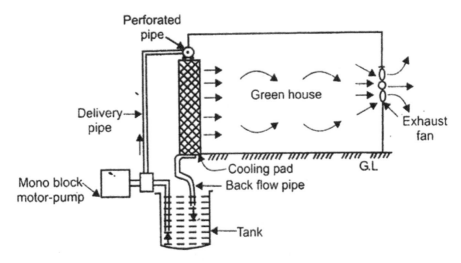

Fig. 3.3: Fan and Pad cooling system
Source: http://eagri.org/eagri50/AENG252/lec04.pdf

2. Fog cooling system

Introduced in greenhouses in 1980, the fog evaporative cooling system works on the same cooling concept as the fan and pad cooling system but uses very different arrangement. Using appropriate nozzles, a high pressure pumping system generates fog which contains water droplets with a mean size of less than 10 microns. These droplets are small enough to remain suspended in air while they evaporate. Fog is distributed around the greenhouse, freshening the air everywhere.

b. Active winter cooling systems

During the winter excess heat can be an issue. In winter, the ambient temperature inside the greenhouse will be below the desired temperature. Due to the greenhouse effect solar heat trapping will increase the temperature to an injurious level; unless the greenhouse is ventilated. The convection tube cooling and horizontal air flow (HAF) ventilator cooling systems are two widely used active winter cooling systems.

1. Convection tube cooling

The general components of the convection tube include the louvered air inlet, a polyethylene convection tube with air distribution holes, a pressurizing ventilator to direct air under pressure into the tube, and an exhaust ventilator to produce vacuum. Once the air temperature inside the greenhouse reaches the set limit, the exhaust fan begins to work thus producing vacuum inside the greenhouse. Due to the vacuum, the louvre of the gable inlet is then opened through which cold air enters. The pressurizing fan works to suck up the cool air that reaches the louvre at the end of the transparent polyethylene convection tube.

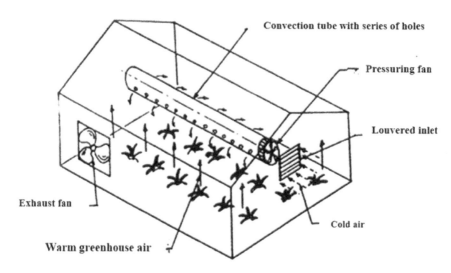

Fig. 3.4: Convection tube cooling system
Source: http://eagri.org/eagri50/AENG252/lec04.pdf

2. Horizontal air flow (HAF) cooling

The HAF cooling system uses small horizontal fans to transfer the air mass, and is considered an alternative to the air distribution convection tube. In this method the greenhouse can be viewed as a large box containing air, and the strategically located fans move the air in a circular pattern. This system should move air to the greenhouse floor area at 0.6 to 0.9 m³/min/m². The fractional horse power of the fans is adequate for operation from 31 to 62 W (1/30 to 1/15 HP) with a blade diameter of 41 cm. The fans should be positioned so as to channel air flows along the length of the greenhouse and parallel to the ground. The fans are located at 0.6 to 0.9 m above plant height and 15 m at intervals. They are arranged in such a way that one row of fans directs the air flow down one side to the opposite end along the length of the greenhouse, and then back along the other side by another row of fans (Fig. 3.5). Greenhouses with larger widths may require more fan rows along their length.

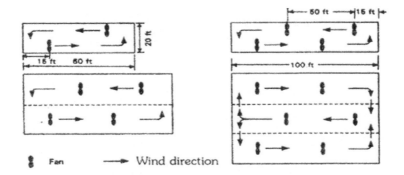

Fig. 3.5: HAF system in different sizes of greenhouses
Source: http://eagri.org/eagri50/AENG252/lec04.pdf

3. Greenhouse ventilation

Ventilation is the process of allowing the fresh air to enter the confined area by driving the unwanted properties out of the air. Ventilation is important in the sense of the greenhouse to reduce temperature, to replenish CO_2 and to regulate relative humidity. Greenhouse ventilation needs differ greatly, depending on the crops grown and the growing season. The ventilation system may be either a passive (natural ventilation) system or an active (forced ventilation) system with fans.

i. Natural ventilation

The sides of greenhouse structures in the tropics are often left open to natural ventilation. Tropical greenhouse is essentially a rain shelter, a polyethylene cover over the crop in order to prevent rainfall from reaching the growing area. That mitigates the foliage disease problem. Ventilators are mounted on both sides of the roof adjacent to the peak, and on both greenhouse side walls. Both the ventilators on the roof and those on the side wall account for about 10 percent of the total roof area each. The south roof ventilator is opened in stages during the cooling process in winter to meet the cooling needs. As the incoming air passes through the greenhouse, it is heated by sunlight and mixed with the warmer greenhouse air. The incoming air becomes lighter with the increase in temperature, and it rises up and flows through the roof ventilators. It produces a chimney effect (Fig. 3.6), which in turn draws a continuous stream of more air from the side ventilators.

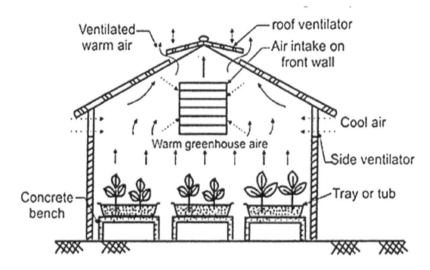

Fig. 3.6: Chimney effect in general passive ventilation
Source: http://eagri.org/eagri50/AENG252/lec04.pdf

a. Roll up side passive ventilation in poly houses

Rolling up ventilation method, allows air to flow through the plants. In response to temperature, prevailing wind and rain, the amount of ventilation on one side or on both sides can be easily changed (Fig. 3.7). During periods of extreme heat, rolling the sides up almost to the top may be appropriate. Passive ventilation can also be achieved by manually lifting or partitioning the sheet of polyethylene.

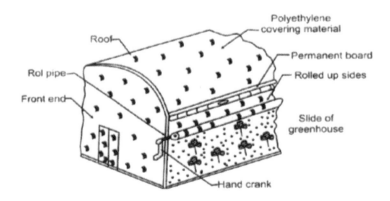

Fig. 3.7: Roll up side passive ventilation
Source: http://eagri.org/eagri50/AENG252/lec04.pdf

ii. Forced ventilation

Mechanical devices such as fans are used to expel the air in forced or active ventilation. Uniform cooling can be accomplished by this method of ventilation. These include cooling systems for summer fan-and-pad and fog, and the winter convection tube and horizontal airflow systems. For mechanical ventilation, the greenhouse ventilation is used for low pressure, medium volume propeller blade fans, both directly connected and belt driven. These are mounted opposite the air intake at the end of the greenhouse, which is usually protected by gravity or motorized louvers. The fans vents, or louvers, should be motorized, with fan movement regulating their motion. Motorized louvers prevent the wind from opening the louvers, especially when the greenhouse is supplied with heat. To prevent hot areas in the crop field, wall vents should be regularly positioned around the end of the greenhouse.

The fan-and-pad cooling system is called evaporative cooling in conjunction with the fans. Typically the fans and pads are arranged in opposite greenhouse walls.

Fogging systems are an alternative to refrigerating evaporative pads. They rely on completely clean water, clear of any soluble salts to prevent the mist nozzles from plugging in. These cooling systems are not as popular as evaporative cooling pads but will be widely adopted as they become more cost-competitive. Fogging systems are the second stage of refrigeration when passive systems are inadequate.

iii. Microprocessors

One may consider dedicated microprocessors as simple computers. A typical microprocessor will have a programming keypad and two or three line liquid crystal display of, at times, 80 character length. We have more connections to the display, and can monitor up to 20 machines. It's cheaper to use a microprocessor with that number of computers. They may receive various types of signals, such as temperature, light intensity, rain and wind speed. We require the integration of the diverse range of devices, which cannot be accomplished with thermostats. The microprocessor's accuracy for temperature control is relatively good. Microprocessors can be made to operate various devices, for example, a microprocessor may control the ventilators for wind direction and speed, based on the sensor details. Likewise, a rain sensor may also enable the ventilators to avoid wetting of the moisture sensitive crop. A microprocessor can be programmed to activate the CO_2 generator when the luminous intensity reaches a fixed point, a minimum photosynthesis level.

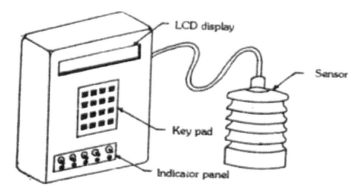

Fig. 3.8: Microprocessor for controlling greenhouse
Source: *http://eagri.org/eagri50/AENG252/lec04.pdf*

iv. Computers

Computer control systems are popular nowadays in greenhouse installations in Europe, Japan and the USA. Computer systems for nearly any size growing facility can provide completely integrated control of temperature, humidity, irrigation and fertilization, CO_2, sun, and shade levels. Precise control over a growing operation helps growers to realize energy, water, chemical and pesticide applications savings of 15 to 50 per cent.

A machine can monitor hundreds of devices inside a greenhouse (devices such as winds, heaters, fans, hot water mixing valves, irrigation valves, curtains and lights, etc.) by using thousands of input parameters such as outdoor and indoor temperatures, humidity, outside wind direction and speed, CO_2 levels and even day or night time. Computer systems receive signals from all sensors, assess all conditions and send appropriate commands every minute to each piece of equipment in the greenhouse range thus maintaining ideal conditions in each of the different independent greenhouse zones defined by the grower (Fig. 3.9).

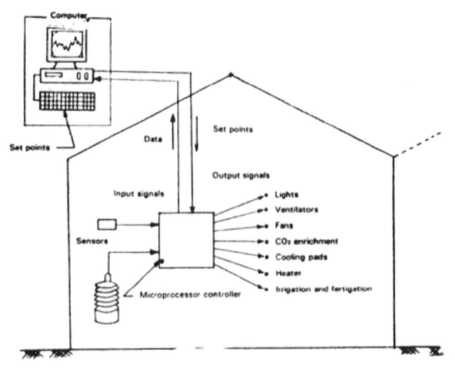

Fig. 3.9: Computerized control systems in greenhouse
Source: http://eagri.org/eagri50/AENG252/lec04.pdf

Advantages

» The machine already knows what all systems are doing and, if correctly programmed, will organize all systems to provide the optimal environment without overlapping.

» The device can capture environmental data that can be displayed to show current conditions or stored and analysed to provide a crop time history and, if desired, can also be shown in a table or graph format.

» A high-speed computer with networking facilities can monitor several remotely distributed greenhouses by placing the device in a central area and monitoring the results regularly.

» The device will predict weather changes and adjust heating and ventilation systems with proper programming and sensing systems, thus saving electricity.

» When conditions are intolerable to and sense sensor and system failure, the device can be programmed to sound an alarm.

Disadvantages

» High initial investment costs.

» This includes professional operators.

» It requires high maintenance, care and precautions.

» Small-scale and seasonal production not economical.

Greenhouse Heating

The northern parts of our country are experiencing cold winters, where heating systems need to be used in the greenhouses along with summer refrigeration systems. Whereas greenhouses in the southern region do need cooling systems, as the cold effect in winter is not that intense. In cold weather conditions, greenhouse heating is needed if the heat trapped during the nights is not adequate.

Modes of heat loss

The first mode of heat is by conduction of materials covered by a greenhouse. Specific materials, such as aluminium sash bars, glass, polyethylene, and cement partition walls, differ in conduction depending on the intensity of heat from the warm interior to the colder exterior. In a shorter time a good heat conductor loses more heat than a weak conductor, and vice versa. There are only limited ways to insulate the coating material without blocking the transmission of light. The optimal device tends to be a dead-air gap between two coverings. A saving of 40 per cent of the heat allowance can be achieved by adding a second cover.

Convection (air infiltration) is a second form of heat-loss. Spaces between glass or FRP (Fibre-reinforced plastic) panels and ventilators and doors allow warm air to flow outwards and cold air inwards. A general assumption is that the volume of air held in a greenhouse can be lost as often as once every 60 minutes in a double layer film plastic or polycarbonate panel greenhouse, every 40 minutes in a FRP or new glass greenhouse, every 30 minutes in an old well-maintained glass greenhouse and every 15 minutes in an old poorly maintained glass greenhouse. Approximately 10 per cent of overall heat loss from a structurally tight glass greenhouse is caused by lack of infiltration.

The third mode of heat loss from the greenhouse is radiation. Warm objects emit radiant energy, which passes through the air to cooler objects without significantly heating the air. The colder objects are getting warmer. Glass, vinyl plastic, FRP and water make radiant energy completely invisible, while polyethylene is not. Polyethylene, greenhouses can lose large quantities of heat from radiation to colder outside objects, unless a layer of moisture forms a shield on the polyethylene.

Heat distribution systems

Heat is spread by one of two typical methods out of the unit heaters. In the convection tube method, warm air from unit heaters is spread via a transparent polyethylene tube that runs through the greenhouse duration. In small jet streams, heat escapes from the tube through holes on either side of the tube, easily interacting with the surrounding air and creating a circulation pattern to eliminate temperature gradients.

The second heat transfer process is by horizontal airflow. The greenhouse can be visualized in this system as a large air-containing structure, and it uses small horizontal fans to transfer the air around. The fans are above the plant height and spaced in two rows about 15 m (50 ft) apart. Their arrangement is such that the heat from one corner of the greenhouse is directed to the opposite end from one side of the greenhouse and then back along the other side of the greenhouse.

Solar heating system

Solar heating is often used as a partial or absolute alternative to heating systems for fossil fuel. Today there are few solar heating systems in greenhouses. Solar heating system's general components (Fig. 3.10) are collector, heat storage facility, exchanger to transfer solar derived heat to greenhouse air, backup heater to take over when solar heating is insufficient and set of controls.

Collecting heat through flat-plate collectors is most efficient when the collector is positioned at solar noon perpendicular to the sun. The heat extracted can provide 20 to 50 per cent of the heat requirement depending on the locations.

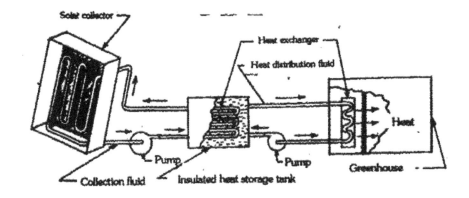

Fig. 3.10: A typical solar heating system for greenhouse
Source: http://ecoursesonline.iasri.res.in

Water and rock storage

Water and rocks are the two most popular sources for greenhouse heat storage. For a 10 °C rise in temperature, one kg of water can hold 4.23 kJ of heat. For every 10 °C rocks can store about 0.83 kJ. A rock bed would have to be three times as large as a water tank to hold equal amount of heat. A water storage system is well suited to a water collector and a greenhouse heating system consisting of a pipe coil or a water coil-containing unit heater. Throughout the day hot collector water is pumped to the storage tank. Hot water is pumped from the storage tank to a hot water or steam boiler or into the hot water coil within a heater unit as and when heat is needed.

Using an air-collector and a forced air heating system, a rock storage bed can be properly used. In this scenario, the collector's heated air is pushed through a rock bed along with overly heated air within the greenhouse during the day (Fig.3.11). Most heat is absorbed in the rocks. The rock bed may be placed under the greenhouse floor or outside the greenhouse, and should be well insulated against loss of heat. When heat is needed in the greenhouse during the night, cool air from the inside of the greenhouse is pushed through the rocks, where it is heated and then moved back into the greenhouse again.

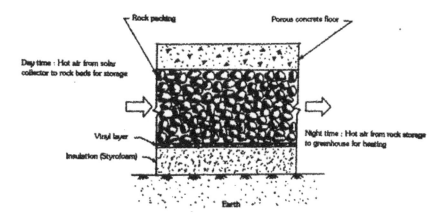

Fig. 3.11: Cross section of a typical rock storage unit
Source: http://ecoursesonline.iasri.res.in

Heat conservation practices

Energy can be saved considerably if the following recommendations from the American Council for Agricultural Science and Technology (CAST, 1975) are applied by a grower:

1. Tighten up the house and close any possible openings.
2. Using polyethylene or fiberglass at the inside of ends of the gable.
3. To avoid leaks maintain the steam or hot water system periodically.
4. To reflect heat into the greenhouse, use reflector materials behind heating pipes.
5. Hold the boiler on peak output.
6. Insulate the supply of hot water and steam and return the piping, and check at intervals, removing the insulation if necessary.
7. Maintain automatic valves inside the heating system.
8. To ensure correct operation, test the thermostats periodically.

The following are the other improved practices for energy conservation:

1. A greenhouse is filled with a double polyethylene film to minimize heat loss.
2. A removable sheet of polyethylene should be positioned over each plant row, and a row cover should be placed over each row so as to minimize

night heat loss.

3. Application of opaque sheets throughout the night as curtains.

4. Application of at least one thermal screen movable layer.

5. Installation of polyethylene tubing, which seals the rising area from the surface of the roof upon installation.

6. Improved curtain materials save more energy with greater reflective properties.

Table 3.4: Greenhouse Accessories

System type	Accessories
Climate Controls	Heating and Cooling, Thermostats, Variable Speed Controls, Humidistats, Cycle Timers
Ventilation	Evaporative Coolers, Exhaust Fans & Shutters, Automatic Vent Openers, Circulation Fans
Heaters	Gas Heaters, Electric Heaters, Heating Mats & Cables
Misting Systems	Sprinkler System, Misting Systems, Mist Timers & Valves, Water Filter
Watering	Plant Watering Systems, Drip Systems, Professional Water Hoses, Watering Timers, Overhead Watering Systems
Meters	Min./Max. Hygro-Thermometer, Min/Max Thermometer, Soil Thermometer, Thermometer & Hygrometer, Light Intensity Meter, pH Meter, EC Meter, LUX meter
Greenhouse Plastics	Shade Fabric, Greenhouse Film, Patching Tape, Batten Tape, Ground Cover Flooring, Greenhouse Bubble Insulation
Grow Lights	CFL Fluorescent Lights, T5 Fluorescent Light Systems, MH & HPS Light Systems, MH & HPS Bulbs & Supplies
Benches & Shelves	Superior Greenhouse Benches
PVC Fittings & Pipe	PVC Fittings, PVC Pipe
Drip line fittings	Grommet, Elbow, Nipple, Joiner, Reducing Tee, Tee, Reducer, End Cap, Lateral Cock
Greenhouse Electrical	Waterproof Outlets, Electrical Wire, Flexible Conduit
Pollination	Electric Pollinator/ Agri. Pollination Tool

Material for construction of traditional and low cost greenhouses

Naturally greenhouse based wood / bamboo falls under conventional and low cost greenhouses. Naturally ventilated greenhouse based on wood / bamboo (also known as rain shelter in heavy rain fall areas) is made from Casurina, Nilgiri, etc. or Bamboo. Preferably, the poles made from these materials can be handled either with turpentine or coal tar at one end, to be mounted in the foundation pits. For fixing the UV stabilized polyethylene films, aluminium profiles with spring or wooden battens can be used. The wooden / bamboo poles and other supporting materials used in the greenhouse construction must be sufficiently strong to withstand different forms of wind load, crop load etc. The recommended outer diameter specifications of wooden / bamboo poles lies between 8 cm-10 cm for main poles and 6 cm-8 cm for purlins & trusses to withstand such loads. All parts of the structure are preferably fitted with the help of the nails or nut-bolts.

The technical specifications suggested for a greenhouse of a wooden or bamboo structure are given below:

Table 3.5: Suggested technical specification for Wooden/ Bamboo based greenhouse

Sr. No.	Item	Description/ Specifications
1	Product	Naturally Ventilated Greenhouse (Wood/ Bamboo based)
2	Size	100 m^2- 250 m^2
3	Width of greenhouse	At least 35% of the desired length of greenhouse
4	Ridge height	3.5 m to 4 m
5	Ridge Vent	80 cm to 90 cm opening fixed with 40 mesh nylon insect screen
6	Gutter height	2.25m – 2.75 m from floor area
7	Gutter slope	The gutter slope should be at least 2 % in foundation structure
8	Structural design	The structural design should be sound enough to withstand wind speed minimum 130 km/hr and minimum load of 20 kg/m^2. There should be provision for opening one portion at either side for entry of small power tiller for intercultural practices.

9	Structure	Complete structures are made of strong Wooden Bamboo posts. The post should have dimension 8 cm- 10 cm diameter for central post, side post and Gutter post/ tie beam etc. And diameter 6 cm-cm for Post Plate, Supporting post, Trusses / members/ sticks / others structural members for joining each other properly.
	Treatment of poles	The post must be treated with different type of preservatives to protect it from termites/ fungal attacks. The recommended preservatives are Coal Tar Creosote, Copper Zinc Naphthenates and Abietates, Boric Acid and Borax, Copper-Chrome –Arsenic (CCA) Composition, Acid- Curpric –Chromate Composition & Copper- Chrome- Boric Composition.
	Fasteners	All nuts & bolts, nails, Aluminium/ MS strip of 2 cm width must be of high tensile strength and galvanized & if required, there should be provision of PP/ Coir & Jute ropes for anchoring the structure.
10	Entrance room & Door	Two entrance door of size 2.5m X 1.5 m must be provided as per the requirements and covered with 200 micron UV stabilized transparent plastic film.
11	Cladding material	UV stabilized 200 micron transparent Plastic films conforming Indian Standards (IS15827:2009), multi-layered, anti-drip anti fog, anti-sulphur, diffused, clear and having minimum 85% level of light transmittance.
12	Fixing of cladding materials	All ends/ joints of plastic film need be fixed with two way aluminium profile with suitable locking arrangement along with curtain top.
13	Spring Insert	Zigzag high carbon steel with spring action wire of 2-3 mm diameter must be inserted to fix shade net into Aluminium Profile.
14	Curtains and insect screen	Roll up UV stabilized 200 micron transparent plastic film as curtains need be provided up to recommended gutter height on all sides having provision for manual opening and closing of curtains should also be provided. 40 mesh nylon insect proof nets (UV stabilized) of equivalent size need to be fixed inside the curtains. Anti flapping strips is suggested to ensure smooth functioning of the curtain.

15	Shade net	UV stabilized 50% shading net with manually operated expanding and retracting mechanism. Size of net should be equal to floor area of the greenhouse.
16	Footpath	1 m wide and 10 cm thick footpaths should be provided in the centre (Length X width) & made of cement concrete 1:2:4.
After sales services		
17	Warranty	The firm to provide Warranty free maintenance for one year from the date of installation.
18	Testing	All plastic materials used in the greenhouse should be tested by the CIPET for quality assurance (If necessary).
19	Training	Free training for operation, maintenance & production for one year.

Chapter - 4

Greenhouse Irrigation Systems

IRRIGATION SYSTEMS

Greenhouse crops are irrigated by applying water to the media surface through drip tubes or tapes, by using a manual hose, overhead sprinklers and booms, or by applying water through the bottom of the container by subirrigation, or by using a combination of these delivery systems. Overhead sprinklers and hand watering continue to "waste" water, and often damp the vegetation, which increases disease and injury risk. Drip and subirrigation systems are the most efficient and give greater control over the amount of water that is applied.

There is also a reduced potential for disease and injury as the foliage does not get wet.

Rules of Watering

The following are the three important rules of application of irrigation.

Rule 1: Use a well-drained substratum with excellent structure.

If the root layer is not well drained and aerated it is difficult to achieve adequate watering. Thus substrates with ample retention of moisture along with good aeration are indispensable for proper plant growth. It is possible to achieve the desired combination of coarse texture and highly stable structure from the formulated substrates and not from field soil alone.

Rule 2: Water thoroughly each time.

Partial watering of the substrates should be avoided; in the case of containers, the supplied water should flow from the bottom, and in the case of beds the root zone is thoroughly wetted. As a rule, excess water is supplied by 10 to 15 per cent. In general, the soil-based substrates need water at a rate of 20 l / m² of bench, 0.3 to 0.35 liters per 16.5 cm (6.5 in) pot diameter.

Rule 3: Water just before initial moisture stress occurs.

Because overwatering inhibits the production of aeration and root, water should be applied just before the plant enters the early water stress symptoms. It is possible to use foliar symptoms such as texture, color and turbidity to assess moisture tension, but they differ with crops. For crops which show no symptoms, the color, feel and weight of the substrates are used for evaluation.

Types of Watering

Hand watering

The most popular irrigation method is hand watering, and it is not economical in the present days. Growers can only afford hand watering where a crop is still at high density, such as in seed beds, or where it is watered in a few chosen pots or areas that have dried earlier than others. The saved labour would pay for the automated device in less than a year in all situations. It will soon become apparent that this is too high a cost. Despite this hand watering deterrent, there is a great danger of adding too little water or taking too long between the watering. Hand watering takes a considerable amount of time, and is very boring. This is normally accomplished by novice workers who may be tempted to rush the job or push this off to another time. Automatic watering is fast and simple, and is carried out by the grower himself. Where hand watering is done, a water breaker at the end of the hose should be used. Such a device breaks water force, allowing for a higher flow rate without washing the root substratum out of the bench or pot. It also decreases the risk of disrupting substrate surface structure.

Perimeter watering

Perimeter watering system may be installed in benches or beds to grow crops. A typical device consists of a plastic pipe around a bench perimeter with nozzles that spray water over the surface of the substratum below the vegetation (Fig. 4.1).

It is possible to use either polyethylene, or PVC pipe. While PVC pipe has the advantage of being very stationary, if not firmly anchored to the side of the bench, polyethylene pipe tends to roll over. This causes nozzles to rise or fall from their proper orientation towards the surface of the substratum. Nozzles are made of nylon or hard plastic and are ideal for putting out a 180°, 90° or 45° spray arc. They are spread across the benches, regardless of the types of nozzles used, so that each nozzle projects out between two other nozzles on the opposite side. Perimeter watering systems with 180° nozzles require up to 30.5 m (100 ft) in length for benches with one water valve. A water should be mounted on each side for benches over 30.5 m (100 ft) and up to 60.1 m (200 ft), one serving each half of the bench. This system is for 1.25 l / min / m tubing. Where there are alternating 180° and 90° or 45° nozzles, the length of a bench serviced by one water valve should not exceed 23 m (75 ft).

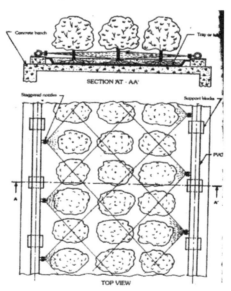

Fig. 4.1: Schematic diagram of perimeter watering system

Overhead sprinklers

Although for disease control purposes the foliage on most crops should be kept dry, a few crops tolerate wet foliage. Most easily and cheaply these few crops can be irrigated from the overhead. Bedding plants, azalea liners, and some green plants are commonly watered crops from above (Figure 4.2). A pipe is mounted along the center of a bed. Riser pipes are periodically installed at a height well above the crop's final height (Fig. 4.2). For bedding plants flats a total height of 0.6 m (2 ft) is appropriate, and for fresh flowers 1.8 m. Each riser has a nozzle installed at the top. Often the trays are put under pots to catch water that would otherwise fall

48

between pots and wasted on the table. Each tray is square, and it meets the tray next door. Quick all the water is intercepted in this way. That tray has a depression that accommodates the pot, and is then angled upward from the pot to the edge of the tray. The trays also have drain holes that allow excess water to drain and store some quantity, which is then absorbed by the substrate.

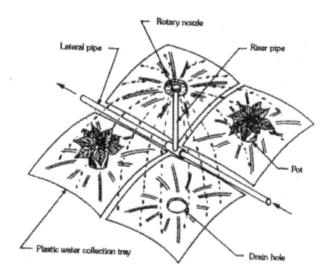

Fig. 4.2: Schematic diagram of overhead sprinkler watering system

Boom watering

Boom watering may act as an open or closed device, and is often used to produce seedlings that are grown in plug trays. Plug trays are plastic trays of approximately 30 x 61 cm (12 x 24 in) in width and length, 13 to 38 mm (0.5 to 1.5 in) in depth, and contain approximately 100 to 800 cells. Each seedling is grown in its own cell. Watering accuracy is extremely important for plug seedlings during the 2 to 8 week development period.

A boom watering system usually consists of a boom of water pipes extending from one side of a greenhouse bay to the other. The pipe has nozzles which can pump either water or fertilizer solution down onto the crop. At its centre point the boom is connected to a carriage that runs along tracks, mostly suspended above the greenhouse bay's centre walk. In this way, the boom can pass from one end of the bay to the other. The boom is powered by an electric engine. The quantity of water delivered per plant area unit is determined by the velocity the boom moves at.

Drip Irrigation

Drip irrigation, also referred to as trickle irrigation, consists of the placing of small diameter plastic tubes on the surface or subsurface of the field or greenhouse next to or under the plants. Water is delivered to the plants through small holes or emitters located along the tube at frequent intervals. Drip irrigation systems are commonly used as an integral and essential part of the comprehensive architecture, in conjunction with safe agriculture. Drip irrigation is the only way to apply even water and fertilizer to the plants when using plastic mulches, row covers, or greenhouses. Drip irrigation provides full control over environmental variability; ensures optimal productivity with minimum water usage while maintaining nutrients from soil and fertilizer; and regulates the cost of water, fertilizer, labour and machinery. The easiest way to save water is by drip irrigation. The distribution capacity is typically 90 to 95 percent, compared to 70 percent sprinkler and 60 to 80 percent furrow irrigation, depending on soil type, field level and how water is applied to the furrows. Drip irrigation is recommended not only for protected farming but also for open field crop production, particularly in arid and semi-arid regions of the world.

Drip irrigation replaces surface irrigation where water is scarce or costly, where the soil is too porous or too impervious to irrigate gravity, land levelling is difficult or very costly, water quality is low, the environment is too windy for sprinkler irrigation, and where skilled irrigation labour is not available or expensive. In drip irrigation weed growth is reduced, as irrigation water is applied directly to the plant row and not to the whole field as with irrigation by sprinkler, furrow, or flood. Placing water in the plant row increases the efficiency of the fertilizer since it is injected into irrigation water and applied directly to the root zone. Diseases of plant foliage can be reduced as the leaves are not wetted during irrigation.

One of the disadvantages of drip irrigation is the initial cost of equipment per acre which can be higher than other irrigation systems. These costs must, however, be assessed by comparing them with the expense of land preparation and maintenance often required by surface irrigation. Basic irrigation equipment consists of a lateral or emitter pump, main line, delivery pipes, manifold, and drip tape, as shown in Figure 14.3 & 15.

Usually, the head between the pump and the pipeline network consists of control valves, couplings, filters, time clocks, fertilizer injectors, pressure controls, flow meters, and gauges. Since the water in emitters passes through very small outlets, it is absolutely necessary that it be screened, filtered, or both before it is distributed in the pipe system. The initial field positioning and architecture of a drip network is affected by terrestrial topography and the expense of various system configurations. Additionally, design considerations should include the relationship between the different system components and the farm equipment required to plant, produce, sustain and harvest the crop.

Drip irrigation can be a valuable method for the accurate regulation of medium rising humidity. This also saves energy and resources, and reduces groundwater contamination capacity. Drip irrigation systems remove runoff of water that is lost during overhead irrigation, and regulate the amount of water that is added to the bowl. In principle, leaching from pots can be significantly reduced or removed by simply turning off the system as container capacity is attained. Controlling drip systems using a tensiometer positioned in the rising medium to sense moisture tension (level) and a small computer programmed to switch on or off the device when present moisture pressures are reached has been shown to minimize runoff from potted chrysanthemums and poinsettias to almost nil.

Vegetable crops are commonly watered with drip bands when grown in ground beds, bags, or pots. Tubing is put through the bags atop the ground or tub, or woven.

Equipments required for drip irrigation system include

- » Pump unit to generate a pressure of 2.8 kg / cm^2
- » Air filtration device-filters for sand / silica / screen
- » Dripper or Emitter PVC tubing

Drippers of different types are available

- » Labyrinth drippers
- » Turbo drippers
- » Pressure compensating drippers – contain silicon membrane which assures uniform flow rate for years
- » Button drippers- easy and simple to clean

Depending upon the type of water, different kinds of filters can be used:

Gravel filter: Used for filtration of water obtained for open canals and reservoirs that are contaminated by organic impurities, algae etc. The filtering is done by beds of basalt or quartz.

Hydrocyclone: Used to filter well or river water that carries sand particles.

Disc filters: Used to remove fine particles suspended in water

Screen filters: Stainless steel screen of 120 mesh (0.13mm) size. This is used for second stage filtration of irrigation water.

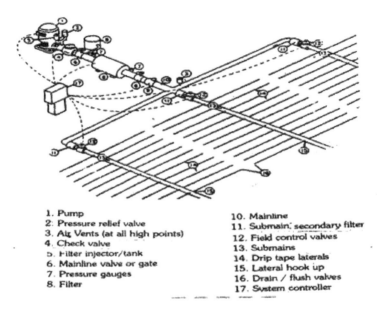

1. Pump
2. Pressure relief valve
3. Air Vents (at all high points)
4. Check valve
5. Filter injector/tank
6. Mainline valve or gate
7. Pressure gauges
8. Filter
10. Mainline
11. Submain; secondary filter
12. Field control valves
13. Submains
14. Drip tape laterals
15. Lateral hook up
16. Drain / flush valves
17. System controller

Fig. 4.3: A typical layout of drip irrigation system

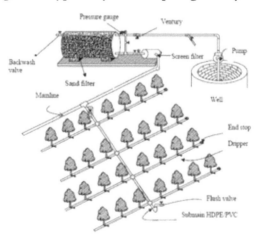

Fig. 4.4: Diagram of drip irrigation system for greenhouse

Source: eagri.org/eagri50/AGRO103/lec15.pdf

Water trays and saucers

Throughout this method, water is added to the surface and collected under the container by means of collection trays or saucers Depending on their shape and placement on the bench, water trays and saucers can significantly minimize leakage and leaching by collecting the water flowing from the pots and retaining the water that escapes the pot during overhead watering. They're cheap, and reusable. Water which collects in them should be given sufficient time before further irrigation to evaporate or be absorbed by the plant.

Avoid close spacing of the plants and inadequate ventilation when using this method to avoid disease problems.

Subirrigation

Subirrigation systems, also known as zero runoff, preserve water and fertilizers as an environmentally responsible alternative. Greenhouse growers are installing them to improve product quality, achieve more uniform growth and increase efficiency in the production process. Water and nutrient solution given at the base of the container in subirrigation systems increases by capillary action through holes in the bottom and is absorbed by the through media. These systems can be adapted to crops cultivated in pots or flats.

Advantages of subirrigation systems

- » Water solutions and nutrients are included and recycled.
- » Over conventional systems, water and fertilizer use decreases by at least 50 percent.
- » Consistent watering of all containers.
- » The size and location of the pot can be adjusted easily.
- » More vigorous growth of plants.
- » The leaves stay dry.

Classification of Subirrigation Systems

a. Capillary mat systems

The pots are placed on a mat in a capillary mat system which is kept continuously humid with a nutrient solution. Various styles of fabric mats from 1⁄4 "to 1⁄2" thick are available. The pots at the bottom take up the solution through holes. The mat is placed over a sheet of plastic on a level table. Water comes from drip tubes that are laid on top of the fabric.

A sheet of perforated plastic film is often applied over the top of the mat to keep the algae under control. Algicides are available, too. When a new crop is started some farmers turn the mat over. Containers that contain nutrient solution and piping should be covered in black plastic or painted black to prevent the development of light and algae.

b. Trough system

In this system, plastic or metal troughs are placed on existing benches, or the greenhouse structure supports overhead. The troughs are mounted from one end to the other, at a slight slope (3" to 6" per 100'). Pot lined up around the trough. Supplied from spaghetti tubes, the nutrient solution is pumped to the high end, flows past the base of the pots and is collected at the bottom end of a cross gutter. The solution returns to be recycled to a storage tank beneath or below the benches.

One advantage of this system over other ebb and flow systems is the circulation of air that takes place between the troughs. Another is the ability to space troughs for pots of varying size. Dry systems tend to be cheaper than bench systems and can easily be installed in existing greenhouses.

c. Ebb and flood benches and movable trays

This device uses watertight benches of 4 'to 6' wide, or mobile watertight trays to hold the nutrient solution. The benches are built perfectly level, typically of plastic or fiberglass construction, to ensure a uniform depth of liquid. These can be mounted either as fixed or as movable depending on the crops to be cultivated. Channels in the bottom of the bench allow the water to be uniformly distributed and to drain quickly when the water supply is shut down. This helps the bench top to dry, reducing the potential for algae growth and disease.

In service, nutrient solution is pumped into the bench from a holding tank to a depth of 3⁄4" to 1" and kept there for 10 minutes or long enough to absorb the solution by the media in the bottle. A valve is then opened, and the liquid is drained back into the tank rapidly by gravity. Low cost PVC pipe is used, because the fertilizer in the water does not impact it. A filter does away with any solid matter. Typically situated in the floor below the benches, the holding tank will have a capacity of about 1⁄2 gallon / sq. ft. of bench space.

The nutrient solution is used again, but when water is applied, changes may have to be made in pH and soluble salts. Many farmers use water treatment with chlorine, light ultra violet (UV), or ozone to avoid disease. Nutrient and flow control can be manual or with a controller. Watering may occur once or twice a week to multiple times a day, depending on the environment and crop size.

Flood floors

Flooded floors work on the same principle and equipped with the same flow and ebb benches. For the floor surface a watertight concrete is needed and it has to be installed as smoothly as possible to avoid pockets. With a great slope, normally 1/4" in 10', a laser transit is used. A concrete contractor who has flood floor system expertise will be recruited. For creating areas, Berms can be built in gutter-connected houses at the post line. PVC pipe with slots or holes is normally installed in the center of the bay floor to supply and remove the nutrient solution as soon as possible.

Large holding tanks are needed, typically made of concrete and lined with plastic, or epoxy paint coated. A 21 'x 200' bay would usually need from 2000 to 3000 gallons of solution. The tank must be large enough in larger greenhouses to hold the liquid from several bays which are operated as one zone. New flood flooring will report high alkalinity as bicarbonates dissolve in the floor.

PVC piping is used to transport the nutrient solution to fertilizers, as it is inert. The nutrient solution is monitored by a computer. To maintain the required nutrient level, fertilizer is applied, typically as individual elements.

Best results are obtained with installation of a floor heating system. This provides consistent heat in the region of the root zone and dries the floor rapidly after the solution has been drained to reduce the potential for algae production and less disease. A system of horizontal circulation of air flow (HAF) can minimize the humidity in the plant foliage. A fork lift carriage and spacing machine could be used to save handling labour.

Tube watering system:

This machine is used mostly for pot watering. The polyethylene micro-tube carries water to every bowl. Such micro tubes are available in different inner diameters that range from 0.9, 1.1, 1.3, 1.5, 1.9 mm and above.

The number of pots that can be watered from a single 19 mm main water pipe depends on the inner diameter of the micro tube used, e.g. 0.9, 1.1, 1.3, 1.5 mm micro tube, 600, 900, 700 and 400 pots can be handled respectively. Such micro tubes will have a weight at the end of the tube so that water speed can be increased, otherwise there is a danger that the tube will be thrown out of the pot and dig a small pit in the medium.

Fertigation System

An automatic mixing and dispensing unit is installed in the fertigation system which consists of three system pumps and one supplying device. The fertilizers are separately dissolved in tanks and mixed in a given ratio and delivered to the plants via drippers.

Fertilizers

Dosage of fertilizers may rely on growing medium. Soilless mixes have lower capacity to retain nutrients and thus need more regular application of the fertilizer. The basic elements are in the pH range of 5.5 to 6.5 at their optimum availability. Micro elements are typically more readily available at lower pH levels while macro elements are more readily available at pH 6 and higher.

Forms of inorganic fertilizers

In greenhouses, dry fertilizers, slow release fertilizer and liquid fertilizer are commonly used.

Slow release fertilizer

During several months they release the nutrient into the medium. Those granules of fertilizer are filled with porous plastic. Once the granules are moistened the inside fertilizer is gradually released into the root medium. A significant thing to bear in mind with regard to these fertilizers is that before steaming or heating of media they should never be applied to the soil media. Heating melts the plastic coating and immediately releases all the fertilizer into the root medium. The high acidity will incinerate the root zone.

Liquid fertilizer

These are water soluble to 100 per cent. These are coming in powdered form. This can either be a single nutrient, or a whole fertilizer. You have to dissolve them in warm water.

Fertilizers suitable for fertigation

A variety of soluble fertilisers have been specially developed for fertigation. Few soluble fertilisers have properties that are ideal for particular soil conditions, while others can typically be used for different soil types. For eg, certain soils have an overabundance of sulphur but may require additional nutrients such as potassium, calcium and/or magnesium. However, acidic soils require potassium, calcium and magnesium and can thus limit the use of acidifying fertilisers.

Nitrogen Sources

Nitrogen is the primary element used in any form of manufacturing, including those used in greenhouses, as it is needed by plants in large amounts, besides being highly mobile through various phases of biogeochemical cycles. Nitrogen is used in the production of various materials and in different ways. Urea and urea ammonium nitrate solutions are known to be the most prevalent sources of nitrogen used as fertilisers. Soluble urea phosphate has also been available on the market today. Fertigation by drip or sprinklers should prevent the use of a free or anhydrous ammonia (a water-free compound).

Major sources of nitrogen, along with details on their use in the fertigation process, are given below.

» **Ammonium phosphate**

It can contribute to a lowering of ph and soil acidification. High calcium or magnesium in irrigation water creates precipitate deposits which can choke drip emitters and drip lines.

» **Ammonium sulphate**

That is a widely used fertiliser. It is an inorganic soil supplement that benefits particularly in alkaline soils. The active ingredients are nitrogen and sulphur. It dissolves easily in water and is convenient for fertigation. It tends to be acidic, which could be a disadvantage if the greenhouse media is acidic.

» **Ammonium thio-sulphate**

It is used both as a fertiliser and as an acidulating (which makes it mildly acidic) agent. As ammonium thiosulphate is added to the soil by fertigation, the sulphur-oxidising bacteria, *thiobacillus spp.*, Oxidises free sulphur to form sulphuric acid. Sulfuric acid dissolves lime in the soil and forms gypsum. Gypsum makes it smoother and helps to preserve healthy porosity and aeration.

> » **Calcium ammonium nitrate**

It is high in fast-acting nitrate-nitrogen, low in long-lasting ammonium nitrogen, and provides calcium. Calcium ammonium nitrate can be mixed with ammonium nitrate, magnesium nitrate, potassium nitrate and potassium chloride.

> » **Calcium nitrate**

It is soluble in water and induces only a small change in soil or water ph. However, if the water is high in bicarbonate, the calcium content can contribute to precipitation of calcium carbonate (calcium).

> » **Urea ammonium nitrate**

Nitrogen is available in three forms-nitrate nitrogen, urea nitrogen and ammonium nitrogen. The nitrate portion is immediately available as soon as it reaches the root zone. The part of urea travels freely with soil water until it is hydrolyzed by the urease enzyme responsible for the production of ammonium nitrogen.

> » **Urea sulphuric acid**

It's very well suited for fertigation. Urea sulphuric acid is an acidic fertiliser that mixes urea and sulphuric acid. The nitrogen and sulfuric acid content of these products differ based on their particular composition. The benefits of this mix alone excludes the drawbacks of their use. Sulfuric acid eliminates the degradation of soil ammonia due to volatilization.

Phosphorus Sources

Monoammonium phosphate, di-ammonium phosphate, monobasic potassium phosphate, ammonium polyphosphate, urea phosphate and phosphoric acid are some of the most common phosphate derived water-soluble fertilisers. However, when applied at high amounts of calcium or magnesium, precipitation and choking of drip pipes or emitters can be induced. The precipitates so formed in drip pipes are quite stubborn and do not dissolve easily. The use of phosphoric acid injection, which also lowers the pH of the irrigation water, is required to clean such drip pipes and remove precipitates. Its usage could be advisable only if the pH of the fertiliser-irrigation water mixture remains low. However, when the pH is high (due to dilution with irrigation water) the phosphate can precipitate due to the presence of calcium and magnesium. One solution that is often effective is to substitute phosphoric acid injections with sulphuric or urea sulphuric acid to ensure that the pH of the irrigation water stays strong.

» **Ammonium nitrate**

It is a liquid fertiliser primarily used as a nitrogen source in greenhouses. It is present in two types of nitrogen: nitrate-nitrogen (mobile and readily available) and ammonium-nitrogen (longer-lasting, because it is converted to nitrate by micro-organisms).

The main sources of phosphorus, along with details on their use in the fertigation process, are as follows.

» **Ammonium polyphosphate**

It can be used as a fertiliser only through low injection levels. If the water being used has high buffering potential (high carbonate / bicarbonate content usually with high pH, i.e., > 8.0) along with a high calcium and/or magnesium content, possibilities of precipitation in drips are very high.

» **Diammonium phosphate (DAP)**

It is one of the most common fertilisers as a source of phosphorus and is fully soluble in water. DAP is a help in high alkalinity conditions, and many greenhouses are facing this issue.

» **Mono-Ammonium phosphate (MAP)**

It is also fully soluble in water and is a decent source of phosphorus and some nitrogen for plants. It contains nitrogen in ammonia sources that are readily consumed by plants.

» **Monobasic potassium phosphate**

Also known as monopotassium phosphate, it contains a large proportion of phosphorus and potassium.

» **Phosphoric acid**

It can be used in a variety of nitrogen, phosphorus and potassium combination formulations. But it cannot be combined with any high-calcium fertiliser. Being a good supply of phosphorus, it offers an added benefit by keeping the pH of the input injections down and helping to prevent precipitation.

» **Urea phosphate**

It is a good supply of both phosphorus and nitrogen. It supplies nitrogen in the form of urea. It is primarily acidic in nature and extremely suitable for acidifying water and soil.

Potassium Sources

Most potassium fertilisers are water soluble and the use of potassium by drip irrigation systems has been very effective. The most popular drawback is that potassium injection contributes to the creation of solid precipitants in the supply tank when potassium is combined with other fertilisers. Potassium sources most widely found in drip irrigation systems are potassium chloride (KCL) and potassium nitrate (KNO_3). Potassium phosphates are prevented for injection in drip irrigation systems. Major sources of potassium, along with their use in the finished process, are provided below.

» **Potassium chloride**

Potassium is complemented by the use of potassium chloride, which is highly soluble and affordable.

» **Potassium nitrate**

It's expensive, but it supplies both nitrogen and potassium simultaneously. Potassium nitrate is approved for use in irrigation water where salinity issues occur since it has a low salt index.

» **Potassium sulphate**

It can be conveniently used in place of potassium chloride in high-saline areas and at the same time offers a source of sulphur as needed in the fertility or soil management programme.

» **Potassium Thio-sulphate (KTS)**

Two types of potassium thio-sulfate are available and are neutral to pure, chloride-free, transparent liquid solution. It is combined with other fertilisers, but KTS blended cannot be acidified below pH 6.0. The proper order of mixing is to first pour water, then pesticide (if any) and then KTS and/or other fertilisers.

Table 4.1: Composition of major nutrients in different Fertilizers commonly recommended for fertigation

S. No.	Fertilizer	N-P-K
1	Urea	46-0-0
2	Ammonium Nitrate	34-0-0
3	Ammonium Sulphate	21-0-0
4	Calcium Nitrate	16-0-0
5	Magnesium Nitrate	11-0-0
6	Urea Ammonium Nitrate	32-0-0
7	Potassium Nitrate	13-0-46
8	Mono-Ammonium Phosphate (MAP)	12-61-0
9	Potassium Chloride	0-0-60
10	Potassium Nitrate	13-0-46
11	Potassium Sulphate	0-0-50
12	Potassium Thiosulphate	0-0-25
13	Monobasic Potassium Phosphate (MKP)	0-52-0
14	Phosphoric Acid	0-52-0
15	NPK	19-19-19 20-20-20

Table 4.2: Solubility of Nitrogenous Fertilizers

S. No.	Types of fertilizer	Nitrogen content (%)	Solubility (gm/litre)
1	Ammonium Sulphate	21	750
2	Urea	46	1100
3	Ammonium Nitrate	34	1920
4	Calcium Nitrate	15.5	1290

Table 4.3: Solubility of Potassic Fertilizers

S. No.	Types of fertilizer	K content (%)	Solubility (gm/litre)
1	Potassium Sulphate	50	110
2	Potassium Chloride	60	340
3	Potassium Nitrate	44	133

Table 4.4: Solubility of Micronutrient Fertilizers

S. No.	Types of fertilizer	Content (%)	Solubility (gm/litre)
1	Solubor	20 B	220
2	Copper Sulphate	25 Cu	320
3	Iron Sulphate	20 Fe	160
4	Magnesium Sulphate	10	710
5	Ammonium Molybdate	54	430
6	Zinc Sulphate	36	965
7	Manganese Sulphatem	27	1050

Compatibility

Mixing solutions of two or more water-soluble fertilisers will often contribute to precipitate formation. Their solutions should then be prepared separately in two different tanks.

Table 4.5: Combined nutrients

S. No.	Fertilizers	Urea	Ammonium Nitrate	Ammonium Sulphate	Calcium Nitrate	Mono Ammonium Phosphate	Mono Potassium Phosphate	Potassium Nitrate
1	Urea		C	C	C	C	C	C
2	Ammonium Nitrate	C		C	C	C	C	C
3	Ammonium Sulphate	C	C		LC	C	C	LC
4	Calcium Nitrate	C	C	LC		NC	NC	C
5	Mono Ammonium Phosphate	C	C	C	NC		C	C
6	Mono Potassium Phosphate	C	C	C	NC	C		C
7	Potassium Nitrate	C	C	L	C	C	C	

Other Macronutrients

Sulphur (S) can also be given as ammonium thio-sulphate, ammonium sulphate or S-sulphate when required. It is ideal for use with urea ammonium nitrate and other soluble fertiliser types for drip finishing. Magnesium sulphate is also used for the processing of magnesium and sulphur.

Micronutrients

They can be conveniently added via the drip system. Copper, iron, manganese and zinc sulphates are extremely water-soluble and pass easily into the drip system. They are quickly oxidised or precipitated in the dirt, and thus their use can be harmful. It is also beneficial to use chelated fertilisers to increase the performance of micronutrient use. Chelate formulations Fertilizers are usually strongly water-soluble and do not choke precipitation drops.

Fertilizer Application Methods

1. Constant feed

High concentration is much safer at all irrigations. This provides continuous nutrient supply to plant growth and results in a steady plant growth. Fertilisation is referred to as fertilization with each watering.

2. Intermittent application

The liquid fertilizer is applied weekly, biweekly, or even monthly at regular intervals. The problem with this is the wide variability of fertilizer supply in the root zone. High fertilizer concentration will be available in the root zone at the time of application, and it is immediately absorbed by the plant. There will be either low or non-existent by the time next application is made. This fluctuation leads to uneven growth rates for plants, even stress and poor crop quality.

Injectors

This system directly injects small amount of concentrated liquid fertilizer into the water lines to fertilize greenhouse crops with every watering.

Multiple injectors

When incompatible fertilizers are to be used for fertilization multiple injectors are needed. Incompatible fertilizers form solid precipitates when mixed together as concentrates. It would adjust the stock solution's nutrient content, which would also block the siphon tube which injector. The problem can be avoided by multiple injectors.

Fertilizer Injectors

Fertilizer injectors are of two basic types: Those that inject concentrated fertilizer into water lines on the basis of the venturi principle and those that inject using positive displacement

A. Venturi principle injectors

1. These injectors work essentially by differentiating the pressure between the irrigation line and the storage tank for fertilizers.

2. The HOZON-proportioner is the most common example.

3. Low pressure, or suction, is generated at the Hozon suction tube opening at the nozzle contact. It draws up the fertilizer from the storage tank and is mixed in with the irrigation water that flows into the Hozon faucet.

4. The mean Hozon proportioner ratio is 1:16. However, Hozon proportioners are not very accurate because the ratio can differ greatly depending on the pressure of the stream.

5. These injectors are cheap and fit for small areas. Because of its small proportion, large quantities of fertilizer application will need huge stock tanks.

B. Positive displacement injectors

1. Such injectors are more costly than forms of Hozon but are very effective in proportioning fertilizer to irrigation lines regardless of water pressure.

2. Such injectors can have a much broader ratio with the most common being the ratio 1:100 and 1:200. Thus stock tanks are of acceptable size for wide areas of operation and these injectors have much higher flow levels.

3. These proportioners power injection by either a water pump or an electric pump.

4. Anderson injectors are very common with single- and multiple-head models in the greenhouse industry.

   a. The ratios vary from 1:100 to 1:1000 with a dial on the pump head to allow flexibility in feeding.

   b. Multihead installations allow simultaneous, unmixed feeding of several fertilizers. This is particularly significant for fertilizers which are incompatible when mixed together in concentrated form (forming precipitates, etc.).

5. Dosatron feature variable ratios (1:50 to 1:500) and a plain water bypass.

6. Plus injectors also feature variable ratios (1:50 to 1:1000) and work as low as 7 GPM at the water pressure.

7. Gewa injectors do inject fertilizer by pressure into the irrigation lines. Inside the metal tank the fertilizer is contained in a rubber bag.

   a. Water pressure causes the fertilizer to reach water supply from the container.

    b.   When loading the containers, care must be taken, because they may break.

    c.   Ratios are variable between 1:15 and 1:300.

8. Unless your injector is mounted directly in a water line, be sure to add a bypass around the injector so that it can accomplish simple water irrigation.

Advantages of fertigation

1. Helps to supply both water and fertiliser simultaneously.

2. Increases yield by 25-30%.

3. Savings of 25–30 per cent of fertilisers.

4. Application and storage of fertilisers in a uniform and effective way.

5. Changes in nutrient needs as per crop.

6. Lower pH can help protect drippers from clogging.

7. Primary and micro nutrients can be provided with irrigation.

8. The appropriate volume of fertilisers can be injected in concentration.

9. Save time, labour and resources.

Points to remember for adopting Fertigation

Gravitational fertiliser tanks or injection pumps such as venturi (a small piece of narrow tube between broader parts for calculating flow or suction) are used to inject the fertiliser as per plant requirements.

1. Pressure-compensating drippers or inline drippers can be used for precision purposes instead of micro tubes.

2. The level of feed depends on the seed, its growth stage and the season.

3. Stock solution should ideally not surpass 10%.

4. The fertiliser solution should be consistent with the other ingredients mentioned in the following session. Compatibility implies the capacity to blend without moisture.

5. Do not use fertilisers in conjunction with pesticides or chlorine.

6. The time required for the supply of fertilisers does not extend the time provided by the supply of water.

Stop excess water source, which can cause fertiliser leaching (drain away from soil).

General problems of fertigation

Nitrogen

Nitrogen appears to build up at the peripheral amount of wetted soil. Therefore only roots alone at the edge of the wetted zone would have ample access to Nitrogen. Leaching and denitrification indicate the nitrogen is lost. Since downward movement results in permanent loss of NO_3-N, higher discharge levels result in lateral N movement and reduces loss by leaching.

Phosphorous

It accumulates near-emitter, and its efficiency is determined by P fixing power. Low pH close to the emitter results in heavy fastening.

Potassium

It travels laterally as well as downwards and doesn't concentrate past emitters. The distribution is uniform than that of N&P.

Micronutrients

Save for boron, if supplied by fertilisation, all micronutrients accumulate near the emitter. Boron gets lost by leaching low in organic matter in a sandy soil. But Fe's chelated micronutrients, Zn may move away from the emitter but not far from the root zone.

Fertigation equipment

Various forms of fertiliser application systems are commercially available by drip irrigation. They're venturi, a nutrient tank, and a piston pump. The selection of a specific finishing system depends on the location, flow, investment ability and precision required. Generally, small cultivators (up to 1008 sq m) use venturi due to lower costs, medium-sized cultivators (from 1008–4000 sq m) use fertiliser tanks or piston pumps, and large cultivators (more than 5 acres) use electrical or automation machinery as the initial investment is quite big. Manual failures in electrical or automation may also be eliminated, in addition to offering ease of service. In general, the control of nutrients during each irrigation is important.

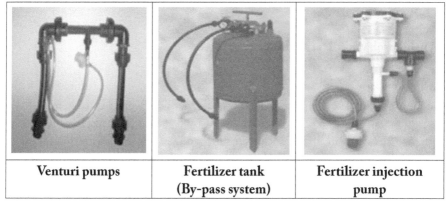

| Venturi pumps | Fertilizer tank (By-pass system) | Fertilizer injection pump |

Fig. 4.6: Different equipments used for fertigation

Chapter - 5

Passive Solar Greenhouse

WHAT IS A PASSIVE SOLAR GREENHOUSE?

Solar energy is stored in all greenhouses. Solar greenhouses are designed not only to collect solar energy during sunny days but also to store heat for nighttime use or during cloudy periods. They can either stand alone or hang in houses or barns.

Passive solar greenhouses are also good options for smallholder farmers because they are a cost-effective way for farmers to prolong the growing season, probably all year.

Basic Principles of Solar Greenhouse Design

Solar greenhouses vary in the following four ways from traditional greenhouses; Sun greenhouses:

» Have glazing oriented during the winter to receive maximum solar heat.

» To retain solar heat, use heat storage materials.

» Have large amounts of insulation where the direct sunlight is low or no.

» Use glazing material and methods for building glazing to reduce heat loss.

» The primary emphasis is on natural ventilation for summer refrigeration.

Solar Heat Absorption

The two most critical factors which affect a greenhouse's amount of solar heat are:

» The greenhouse position or location in relation to the sun

» The type of material used for glazing

Solar Orientation

Given that the energy of the sun is strongest on a building's southern side, glazing for solar greenhouses should ideally be facing the true south. However, if trees, mountains, or other buildings obstruct the sun's path while the greenhouse is in a true south orientation, a true south orientation within 15° to 20° will provide around 90 percent of the true south orientation solar capture. Also, the latitude of your position and the position of possible obstructions that allow you to change your greenhouse orientation slightly from the South to achieve maximum solar energy gain.

Slope of Glazing Material

In addition to the orientation north-south, greenhouse glazing should be ideally sloped to absorb the maximum amount of the heat from the sun. A good thumb rule for having the right angle is to add 10° or 15° to the latitude of the location. For example, if you are at latitude 40° North in northern California or central Illinois, the glazing should be sloped at an angle of 50° to 55° (40° + 10° or 15°).

Glazing

Glazing materials used in solar greenhouses should be able to penetrate the greenhouse with the largest amount of solar energy while reducing energy loss. Rough surface glass, rigid plastic double-layer, and diffuse fiberglass light, while transparent glass transmits direct light. Although plants grow well with both direct and diffuse light, the direct light causes more shadows and uneven plant growth through glazing subdivided by structural supports. Diffuse light passing through glazing evens the shadows created by structural supports, contributing to more cancel growth of the plants.

A solar greenhouse will, as a general rule, have about 0.75 to 1.5 sq. ft. Glazing the floor space per square foot.

Solar Heat Storage

In order to keep the solar greenhouses warm during cool nights or on cloudy days, solar heat entering on sunny days must be stored in the greenhouse for later use. Putting rocks, concrete, or water in direct line with the sunlight to absorb its heat is the most common method for storing solar energy.

Cob, strawbale, stone or cinder block walls at the greenhouse's back (north side) may also provide heat storage. However, only the outer four inches thickness of this storage material effectively absorbs heat. Medium to dark ceramic tile flooring can provide some heat storage, too. Walls not used for heat absorption should be light colored or reflective for directing heat and light back into the greenhouse and providing plants with a more even distribution of light.

Heat storage material (Water is recommended)

The amount of material required for the heat storage depends on your location. You would need at least 2 gallons of water or 80 pounds of rocks to store the heat transmitted through each square foot of glazing if you live in southern or mid-latitude settings. If you live in the northern states, it will take 5 gallons or more of water to absorb the heat entering through each square foot of glazing.

The amount of heat storage material required also depends on whether you are planning to use your solar greenhouse to extend the growing season, or whether you are planting in it all year long. You'll need 2.5 gallons of water per square foot of glazing, or about half of what you'd need for year-round production, for season extension in cold climates.

If you use water as a medium for heat storage, ordinary 55 gallon drums painted a dark, non-reflective color that works well. Smaller containers, such as milk jugs or glass bottles, are more effective than 55-gallon drums in providing heat storage in sometimes cloudy areas. The smaller container has a higher surface area ratio, which results in greater heat absorption when the sun shines. Clear glass containers provide the benefits of absorbing heat better than containers of dark metal and not melting, but they can be broken easily.

Features of a typical passive solar greenhouse

» The Insulated foundation. Unlike other container or hydroponic greenhouses where plants are grown, the passive solar greenhouse enables the gardener to plant directly into the field. For two feet of insulated base, the soil stays warm enough to grow plants year round, depending on the depth of your frost line.

» Glass or glazing made of polycarbonate, both more durable than plastic. Polycarbonate is a hard to crack, lightweight glass substitute that is user-friendly, does not burn the plants, and is assured against yellowing for ten years.

» Super-insulated side walls (either straw bale or conventional insulation) and roof

» Rear north facing straw bale or brick wall

» Passive Solar Panel

Using the sun to power the ventilation system (increased by human powered winds). You can configure the size, location and number of vents to match your heat and humidity profile

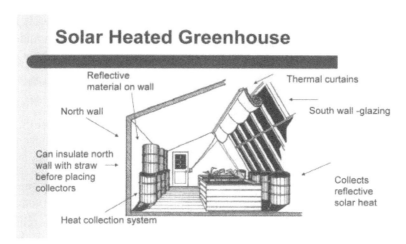

Fig. 5.1: Solar heated greenhouse
Source: https://bradford.missouri.edu

Solar thermal energy is the most plentiful of all the primary, domestic and sustainable or inexhaustible renewable energy options, and is available in both direct and indirect forms. The conversion of solar energy using photovoltaic systems has contributed to speculation on real estate with a concomitant price rise and dire social consequences. Improvements in efficiency through mechanization following a transition in fossil fuel are now negligible, resulting in demand for carbon-free electricity.

Agrivoltaic

Electricity is generated from Photo Voltaic (PV) panels mounted in designed spacing and height, so limited shading allows for productive farming on the ground below. The principle of agrivoltaics is also known in Japan as the "Solar Sharing"

Two main factors for agrivoltaics

1. Through correctly constructed spacing photovoltaic installations sufficient sunlight may hit ground level.

2. Plant species exhibit different shading tolerances and different light saturation point levels.

   » The growth of plants increases with the amount of incident light hitting the light saturation level, and no further increase in light intensity contributes to growth.

   » Excess and unused light may detriment the growth and quality of the plants.

   » Consequently, a well-designed agro-voltaic power plant could generate similar agricultural yields as compared to the open plot and generate electricity as additional power.

Features

» Generating electricity where it is not grid accessible.

» The production of photovoltaic energy has tremendous potential to minimize greenhouse gas emissions.

» Provides additional revenue and work opportunities in rural India;

» Simple light control for shade loving crops.

» Allowing normal tropical growth where seasonal growth is abnormal and dull with high temperatures and strong sunlight.

» Electrification to village.

» The revenue from the PV system is ensured in the event of total crop failure during drought or aberrant weather.

Photovoltaic greenhouse

> » Cohering greenhouse capacity and resource efficiency with the production of renewable energy; photovoltaic greenhouse (PVG) provides a credible solution.

> » Energy is the biggest overhead cost of farm greenhouse production in temperate regions.

> » In integrated photovoltaic greenhouse, socio-economic and environmental efficiency can be higher than separately in the solar or greenhouse system.

> » The use of semi-transparent (photo selective) PV modules for PV greenhouse models seems promising.

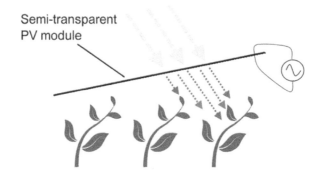

Fig. 5.2: Semi-transparent PV module
Source: Emmott *et. al* (2015)

Types of cells used in preparation of solar modules

1st Generation

Silicon based, such as monocrystalline (c-Si), polycrystalline (p-Si), ribbon crystalline silicon (r-Si), amorphous silicon (a-Si).

2nd Generation

Non-silicon based, such as cadmium telluride (CdTe), copperindium (gallium) and selenide (CIS or CIGS).

3rd Generation

New concept devices, such as concentrated PV (CPV), organic cells (single or tandem), Dye-sensitized cells (Gratzel), Perovskite cells.

Benefits of Agrivoltaic

Agrivoltaic enables the dual use of the same piece of land for electricity generation and food production; the system will also have the following characteristics and benefits, which will contribute significantly to the economic and general development of rural people, particularly farmers.

Table 6.1: Benefits of agrivoltaic

S.No.	Features	Benefits
1.	Growing crops during dry season	The Agrivoltaic system will allow shade-loving crops such as ginger, stevia, gourds, and culinary herbs to grow, thus increasing the land's productivity during the dry months and making it more fertile. Throughout the year, these crops will allow agricultural activity and increase the number of crop cycles to two or more, and will also open up new markets and income for farmers.
2.	Water efficiency	Water that is used to clean the solar plant may be recycled for crop irrigation. Drip irrigation and rainwater harvesting along with water treatment with the reverse osmosis can ensure optimal water use. Air can be supplied with solar energy, too.
3.	Source of employment to local population	In three regions, the agrivoltaic system provides better job opportunities for the local population Maintenance of solar power plants and agricultural activities. It will produce year-round incomes for the local population while improving living standards in the area. Constant electricity availability can help the rural population grow small businesses and manufacturing units that regularly employ rural people. The construction of crop processing and conservation units would help to increase the return of farm production and minimize rural youth migration in town.

Major limitations of Agrivoltaic

» The production of renewable energy requires initial high and large-scale investment.

» Wider adoption of the system is possible in developing countries such as India by promoting conspicuous public subsidies.

» This system may occupy large areas of abandoned farmland.

» Lack of adequate microclimate classification within photovoltaic greenhouses with high shading rates.

Future Prospects

» Photovoltaic system application in solarisation of the soil.

» Solar renewables provides a complimentary solution to many wine growing processes through the Solar Winery model.

» The development of new efficient and cost-effective ways of storing solar energy.

» In order to counterbalance radiation mitigation under photovoltaic greenhouse, suitable species and correct cultivation techniques have to be identified.

» The idea of agrivoltaics is at an early stage of growth, and the performance needs to be closely monitored for sustainability evaluation.

Chapter - 6

Hi-tech Nursery Management

INTRODUCTION

The nursery is a site where the plants are developed and produced to useful sizes. Nursery procedures include the cultivation of seedlings, grafts and grafts of commercially valuable ornamental plants by experimental methods. Many modern techniques are now possible, which are cheap and effective. Such modern methods are helpful in improving the success rate of grafts and cutting rooting; raising the vigour of seedlings; the effect of transplants and generally decreasing the quantity of manual labor. The nursery management acquired the status of a business enterprise where retail nurseries market planting products to the general public, wholesale nurseries that market exclusively to other nurseries and professional landscape gardeners, and private nurseries that serve the needs of establishments or private properties.

The market for high quality planting material is gradually growing due to competition in vegetable farming, fruit tree production, social forestry, agroforestry and planting crops. The need to set up nurseries to fulfil people's needs has been felt by small and marginal growers as well as by gardeners and farm owners. In order to satisfy this need, there is sufficient room for the establishment of small nurseries, which will raise the income of the vulnerable sections of rural society. There is also a requirement for low cost nursery techniques. That kind of nursery decreases physical work and drudgery and improves the vigour of seedlings. This further decreases transplant shocks and the effective rate of grafts and cutting rooting is strong and lower operating costs. There are less supplies, such as energy, fertilizers, etc.

Table 6.1: Disparity between traditional and hi-tech nurseries with respect to equipment and techniques

Conventional nursery	Hi-tech nursery
Spade, khurpi, watering cane, fork, hoe, garden line, roller, basket, sirki, polythyne sheet, sprayer, alkathene sheet, nose-cane, duster, sticks, tags etc.	Plug trays, perforated plastic trays, strip peat pots, nursery stand, sprinklers, protected structures, water pumping motor, media mixture, rakers, temperature control devices, humidity control devices, exhausters, media pressure, seed dibbler, etc.

Although there are limitations in hi-tech nurseries such as greenhouses that are quite expensive and require infrastructural facilities such as large space, water supply and also uninterrupted power supply. The viability of nurseries depends on the conditions of the market. The key criteria for nursery management are as follows:

1. The nursery location should be chosen because there is ample of sunlight and good ventilation.

2. The nursery site will be at a higher position such that water retention can be prevented.

3. In humid and rain-prone regions, nursery regions should be well covered from heavy rains by safe frameworks.

4. The location should be adjacent to and conveniently open to irrigation infrastructure.

5. This will be secured from feral animals, snails, rodents, etc.

6. The soil would be sandy loam or loam with a pH level of 6 to 7 and high in organic matter and safe from pathogenic inoculums.

The Need for a Nursery

Nursery is a must for any horticultural farmer. The production of seedlings in nurseries not only decreases the crop period, but also improves the uniformity of the field and thus improves harvesting relative to direct seed crops. Transplantation of seedlings also eliminates the need for thinning and provides good opportunities for vigorous and off-season virus-free nurseries to grow under protected conditions. Nursery is helpful and convenient to manage small area seedlings and the grower can take timely plant protection measures with minimal effort. The construction of a nursery offers the developing plants a suitable habitat for their improved growth and production. Efficient use of the unfavourable time by the planning of nurseries under protected conditions. Seed costs of certain crops, such as hybrid potatoes, ornamental plants, spices and berries, may be saved by nurseries. Nursery output aims to sustain an efficient plant by gap fillings in the shortest period possible. While several fruits and vegetable seeds can be planted directly in the field, experience has shown that growing seedlings in the nursery has a variety of advantages, as described below.

1. **Intensive care** - Seedlings are provided proper treatment and safety (from livestock, weeds and pests) in the nursery. The average garden soil is not the best way for growing seedlings, particularly from the point of view of soil tilth. At an early stage of growth, most vegetable crops need special care which is not feasible in the main sector.

2. **Reduction of costs** - Fewer seeds are used for growing seedlings in the nursery than for sowing directly in the field, as in the latter seedlings they have to be thinned to one that is unsustainable. Therefore, as costly hybrid seeds are used, transplants are more commercially desirable. Pesticides and labor are often decreased under nursery conditions relative to planting directly in the region.

3. **Opportunity for selection** - The selection of seedlings in the nursery offers the grower the ability to choose well-grown, healthy, standardized and disease-free seedlings.

4. **Extend a short growing season for late maturing crops** - Seedlings can be grown in a nursery in a protected environment until environmental conditions are appropriate for growth and transplanted to the field when conditions permit, thereby minimizing the amount of time spent in the greenhouse.

5. **Forced vegetable production for an early market** - Generally, prices of horticultural products are attractive when production or supply is low. Vegetables may be grown 'out-of-season' in nurseries where conditions are not yet favorable. These crops would therefore grow sooner after transplantation and therefore have a better selling price.

Type of Nurseries

In the past, nurseries have been involved in almost every aspect of plant production and cultivation. They have grown a wide variety of plants and sold them both wholesale and retail, as well as a wide range of allied products and services. Nowadays, all but the largest nurseries prefer to specialize.

Organisational structure

Nurseries can be non-profit organizations (for example, managed by government, civic or environmental groups) or private businesses. Private nurseries may be individually held, a corporation or a business, such as a business with a legal name of its own.

Nurseries may be categorized in a number of ways:

» Depending on whether they produce, such as indigenous, exotics, seedlings, cottages, greenhouse plants, bonsai or bulbs.

» Where they cultivate it, for either in the processing of land or in containers.

» The size of the plants they produce, such as tuber stock, small pots / containers, or advanced stock.

There are three major forms of nursery; growth, increasing and retailing.

Production nurseries (Wholesale nurseries)

Propagate plants and cultivate seedlings and then market them directly to supermarket stores, landscaping and horticulture or forestry offices, or wholesale them to nurseries. Important feature of productive nurseries output.

1. **Innovation** - Supplying new varieties to the market or creating innovative methods of growing and selling established varieties would enable the grower to build new markets.

2. **Specialization** - Increasing fewer lines in greater volumes makes it easier for the grower to increase performance.

3. **Meeting market demands** - Knowing or probable to want market demand and customers' choices, and growing them in sufficient quantities, allows the grower to meet consumer requirements and improve customer loyalty.

Growing-on Nurseries

Growing-up nurseries purchase bulk volumes of seedlings or tiny plants from propagators. During the point of purchasing, the plants are produced in pots, trays or coils, and the plants are then placed into wider containers and maintained for a period of time, contributing interest to the cost of the nursery. In addition to can plant scale, advanced growing methods, such as topiary, can be used to add value to the plant during the growing process.

The most important element of growth in nurseries is the manufacture of a premium commodity for the retail industry. At the time of resale, every plant must be at its peak, showing healthy, vigorous and robust growth.

Retail nurseries

Retail nurseries buy plants from the production / propagation nurseries and resell them in limited quantity as per the needs of the customer. In addition, the store also offers seedlings, seeds, containerized and bare-rooted trees, containers, dried potting mixtures, fertilizers, sprays and bulk landscaping materials.

Nursery Management

The first phase in the effective growth of horticultural crops is the development of good robust seedlings. Young plants, whether propagated from seed or vegetative, need a great deal of care, particularly during the early stages of development. They must be shielded from extreme weather, heavy rains, heat, wind and a broad variety of pests and diseases. When small seeded crops are sown directly in the field, germination is always weak and young plants develop very slowly and take a long time to mature. The season can also be too limited for complete production in the sector. In order to overcome these problems, many vegetable crops are grown in nurseries before being planted in the field. The nursery of crops or fruit is a location where the plants are cared for during the early stages of development, offering optimal conditions for germination and eventual growth before they are adequately healthy to be placed in their permanent position. The nursery can be as simple as a raised bed in an open field or as sophisticated as a glass house with micro-sprinklers and an automatic temperature control system. They also assume that nursery management and plant propagation are the same, although they are entirely separate yet interrelated. Of addition, the mass multiplication of quality planting materials is a central theme of nursery management, but nursery management is a trade-oriented dynamic method, which refers to the effective use of resources for economic returns. Nursery administration is a collaborative endeavour to accomplish the ultimate aim.

The main phases of nursery management are

i.   Planning-edaphoclimatic and socio-economic considerations; demand for planting materials; provision of the mother block, land requirements, water supply, working tools, growing structures and availability of inputs; accessibility; manpower training; plant protection; disposal of planting materials, etc.

ii. Land management, defences against biotic intrusion and soil degradation, correct construction, material supply.

iii. Control and assessment – personal appearance, quick reaction, vital examination, motivation for staff and

iv. Reviews for further refinement. The core elements of hi-tech nursery management are (1) the site, (2) the plant and (3) the individual behind it.

Advantages of nursery management

» It is necessary to have optimal conditions for production , i.e. germination as well as development

» Better care for younger plants as it is easy to take care of nurseries in small areas against pathogenic infections, pests and weeds.

» Crop produced by nursery raising is very early and gets higher prices on the market, rendering it economically more competitive.

» Land management and labor, when the major fields would be filled by crops after 1 month. More intensive crop rotations may be followed.

» There is more room needed for the planning of the primary area as the nursery is developed separately.

» Since vegetable seeds are especially costly hybrids, so we can save the seed by sowing them in the nursery.

DO's and DON'Ts in Nursery Management

DO's

» Adoption of double door system.

» Prefer to solder nuts and bolts instead.

» Work on a sheet of tarpaulins when filling media plug trays.

» Maintain EC and pH control of water and nutrient solution using EC meter and pH meter.

» The seedling trays are wrapped in plastic crates.

» Place plastic crates in the transport vehicle properly.

DON'Ts

» Do not spray when people have no protective clothing to work.

» Do not leave spaces in the nursery.

» Do not dispose of waste close to nursery.

» Do not encourage weeds to grow at the edge of the nursery.

» Do not leave any holes as weed mat spreads, which may contribute to weed emergence.

» Don't let water stagnate in and around a kindergarten.

» Do not leave open the door.

» Don't leave insect holes in the net.

» Do not require the seedlings to touch soil.

» Do not fold in crate seedling trays.

» Don't tightly stack sown seedling trays.

» Do not allow the growth of other plants within the nursery.

» Do not overstretch rising media.

Hi-Tech Nursery Raising

Plug tray nursery raising

A variety of containers may be used to collect vegetable seedlings or nurseries. Plastic trays are used worldwide as regular ones. Plastic trays or depicts that have various cell sizes are usually used to collect seedlings. In general, seedlings are used to lift plastic trays or depicts with various cell sizes. These trays help to germinate properly, provide separate space for each seed to germinate, reduce mortality rates, maintain uniform and safe seedling production, are easy to manage and store, and are durable and economical in transport. Before filling media, these nursery trays may be fixed in thermocol base trays which have the same number and size of cavities. If thermocol base is not available, only trays can be used which are mounted on the floor or firm base.

Root media

Soil was historically recognized as part of the growing mixture of seedlings. The amount of undesirable material, such as diseases, weeds, trash and clay that comes with the soil is high. Soils may contain unknown plant toxins, extremes in variations in soil pH, content of nutrients or unknown chemical residues. Alternatives to the soil are not without difficulties, either in part or in whole. Mostly artificial soilless media are used in the plastic portrayals to collect healthy and vigorous seedlings.

There are various types of soilless media available which can be used to raise plug tray nursery. Below are some soilless materials used as rooting media:

» **Coco peat**: It is completely free of any plague or pathogen infestation. It is widely used as a method to grow vegetable nurseries and ornamental plants.

» **Perlite**: It is composed of light rock of volcanic origin. It is essentially aluminum silicate rock that is spent on heat. It allows for better aeration and drainage. Perlite responds neutrally and gives the media virtually no nutrients. Perlite is used as an inert, light substance to minimize bulk density. It does not retain moisture or hold any nutrient in the plant. As part of rising media there is no real benefit of getting perlite.

» **Vermiculite**: It's mica lost on oil. This is very light in weight and has minerals to enrich the mixture (magnesium and potassium) to enrich the mixture as well as sufficient ability to retain water. Neutral in reaction (pH) according to scale, it is available in grades. By leaching it may reduce the loss of nutrients.

This contains even small quantities of potassium and magnesium.

» **Pine bark:** The effect of its toxins on plant growth and its hydrophobic (or difficult to humidify) nature are two key disadvantages with pine bark. There were also problems with sample purity at times, and with the uniformity of particle size after milling. The bark should be milled to a maximum particle size of 5 mm for the seedling mixes. By aging in open air piles, much of the toxin content can be leached.' but the bark must first be soaked. Newly stripped bark, after drying, is harder to wet than stockpiled, weathered bark. With the introduction of small amounts for slow release fertilizers, pine bark can be composted to aid in the process. This may offer some control over some diseases that may contaminate the bark. The heat produced in the heap will not be even in open stacks, and therefore in the cool areas, the temperature will not rise to many diseases' thermal death point.

» **Hardwood sawdust:** Sawdust issues are similar to those with pine bark, i.e. some timbers have very high levels of toxin. Treatment is similar. The composted sawdust can be mixed with pine bark and peat moss in the prescribed manner. Holding the amount per volume of sawdust to around 40 percent or less is desirable.

» **Soft wood sawdust:** *Pinus radiata* is the most common softwood sawdust available which contains less toxins than some of the Australian hardwoods.

» **Sand:** The purpose of adding sand, an inert mineral, to any seedling mix is to strike a balance with the other components and give the final product some bulk. A popular misconception is that sand actually helps with drainage may happen the opposite. When applied to peat moss, for example, it can fill several of the narrow spaces between peat particles, thus reducing moisture loss. This also increases the re-wetting of a dried-out mix if a small amount of sand is present. Nevertheless, over-mixing of peat and sand in a rotary mixer will cause the peat to pulverize and thus alter its structural characteristics. On the other hand large quantities of sand added to a mixture of seedlings are likely to produce rather dense mixtures. There has to be equilibrium between the poles of fine and coarse sands as well. Classified as coarse material is sand 2 mm or greater.

» **Perlite: Vermiculite:** The nursery sector has long respected horticultural vermiculite. In short-term mixes like those needed for seedlings it has the most value. As a substance with high exchange potential at the heart.

» **Polystyrene pellets or beads:** Often this material is used, if for no other reason than to add volume- It has no nutritional value or ability to retain water. When there is a storm, polystyrene is difficult to treat from outside. Though the results are stronger when mixed dry with a bit of brown coal, before adding moisture.

» **Brown coal:** This substance is also known as coal fines or Lignum Peat. For a soilless mix, the ratio of brown coal to other components must be small, around 20 per cent by volume or even less. Higher volumes with adequate nutrition can result in plants responding with lush growth and hard to harden. Brown coal has a high ability to carry fuel, and a high degree of insufficient water. In addition, its water loss is also high, and it has been shown that there is less water transpiring in brown coal than in peat moss.

» **Peat moss:** Worldwide, the use of peat moss for horticulture is growing. A variety of peat deposits occur in Australia but not all are suitable for seedling mixtures. A typical local peat used in Victoria comes from Tasmania but its quality does not compete with other imported peats due to its trash content (old plant material). Due to their thick nature the fine sedge-peats are usually not ideal for seedling mixtures. Some of the peats are fairly stable against rapid degradation. They are valued for their high water-holding capacity; sphagnum moss have a better water-air balance than most sedge peats. While peats have this water retention ability, their loss of moisture is high through transpiration and evaporation. Because of this it is believed that plants growing in this material will grow faster under optimum conditions.

Sterilisation of Growing Media

Sterilization may be characterised as the process of elimination or degradation of all types of microbial life. In fact, every sterile object in the microbiological sense must be free from any live micro-organisms. Micro-organisms may be destroyed, prevented or eliminated by the exposure of substance to lethal agents that may be infectious, chemical or ionic in nature or, in the case of liquids, by the actual extraction of cells from the medium.

Soil sterilisation

Soil or soilless media are used to raise plants, sustain plants, maintain moisture, and supply the root system with water and nutrients. The media used in the production of plants are often also congenial to the development of micro-organisms. Bacteria, fungi, actinomycets, protozoa, bacteria, worms, nematodes and herb seeds. Microorganisms include both beneficial and detrimental organisms, i.e. soil-borne plant diseases that trigger creatures. In order to remove soil-borne bacteria, nematodes, insects and weeds for safe plant growth, soil or soilless media must be sterilised or pasteurised.

Methods of soil disinfestation

A variety of techniques and agents for soil disinfestation are available. They function in several different forms and each has its own application restrictions. The selection of a system depends on the desired effectiveness, its applicability, safety, availability and cost and the impact on the properties of the item to be disinfested.

Of the number of physical and chemical agents and techniques available, humid heat, i.e. steam sterilisation and chemicals, i.e. fumigants, are most widely used for soil or substrate sterilisation.

Soil solarisation

High-intensity solar radiation during the summer (April – June) is used as a lethal agent to monitor plant pathogenic plants, mosquitoes, nematodes and weeds through the use of translucent polyethylene film and is known as soil solarisation. The step-by - step method for soil solarisation includes

1. First, land should be ploughed.

2. Irrigate the field very gently.

3. Field cover with clear UV-stabilized 25 micron polyfilm for 20–30 days.

4. The sides of the film should be covered with soil to prevent the entry of outside air.

5. Solarisation of soil is not a fool proof form of sterilisation.

Soil sterilisation by formaldehyde

It is an effective sterilising agent for handling toxic soil microbes. It is sold in aqueous solution as formalin containing 37–40 per cent of formaldehyde. The soil or root substrate to be sterilised is loosened and the solution prepared by mixing 4 litres of formalin in 19 litres of water is poured or sprayed into the soil @5 ml / sq m area. The rate of application depends on the moisture content, soil depth and soil form. The land is covered with thin plastic film to contain the fumes emitted.

Removal of plastic film (after 7 days), total evaporation of the scent of formaldehyde can take place in between 15-20 days. After that, it is essential to sow or plant. It has a minimal impact on nematodes and can not be used in standing crops. Its use can ideally be discouraged as it is a general biocide (a material that kills or prevents the growth or operation of living organisms) that is detrimental to the health and welfare of the production system.

Soil sterilisation by hydrogen peroxide

Hydrogen peroxide with nanoparticle silver can be used for sterilisation. Since this solution is in liquid form, it can be applied using a drip irrigation method. The prescribed dosage of the solution is 35–40 ml / sq m, but caution should be taken to ensure that the soil beds are gently watered beforehand. The biggest benefit of having this solution is that sowing / planting can be achieved on the very same day.

Other methods of sterilisation include heat or steam sterilisation, which have minimal use under field conditions due to high expense.

Making a suitable nursery media

For growing nursery, mainly three ingredients-cocopeat, vermiculite and perlite are used as the root medium. Upon filling in plastic portrays these ingredients are combined in a 3:1:1 (Volume Base) ratio. It was also critically noted that safe planting materials can be grown in coco peat alone, which is as good as elevated in three rooting media constituents. Coco peat from Sri Lanka and India contains many nutrients from macro and micro plants including large amounts of potassium, sodium and chloride. So, before using coconut peat as a rooting medium, it is always very important to wash coconut peat 2-3 times with good quality water to eliminate excess of water-soluble elements such as potassium, sodium and chloride. Coco-peat washing helps reduce the electrical conductivity to a tolerance level. Calcium nitrate @ 100 g per 10 liters of water for 5 kg of coco-peat is then used for coconut peat buffering. Calcium [2 +] is added during this cycle to extract monovalent positive ions such as potassium [1 +] from the cocoa complex. In this way we can remove not only water soluble elements but also elements that are bound to the complex of coconuts. Ideally, the treated water should be applied over a 24-hour period in coco-peat through a slow sprinkler system if necessary, otherwise coco-peat can be rested for 24 hours in a calcium nitrate solution. Once the resting period is over, coco-peat is rinsed with water twice, now coco-peat is ready to use as rooting medium.

Selection of a seedling tray

Plastic seedling trays (also known as plug trays or pro-trays) of different sizes are available in the market. Selecting the seedling tray depends on the size of the seeds, trays with 102 or 104 cavities (or plugs) with less cell volume are often used for small seed crops such as tomato, chilli, brinjal etc., while trays with 70 or 50 cavities with more cell volume are used for large seed crops such as most cucurbits, papaya.

Methods of sowing

Manual sowing: Sowing the seeds with the aid of a dibbler that pokes uniform holes in the medium to maintain uniform seeding depth and speed. The minimum dibbling depth is triple the diameter of the seed. A layer of vermiculite or finely sieved coco peat is spread over the sown seeds to a depth 1-1.5 times the seed diameter after sowing. As the seed emerges it will be easier for the epicotyl to drive up through the light media.

Machine sowing: A seeder assembly can be used in large scale nurseries for the automatic loading, sowing, covering and watering of trays. It allows for filling about 90-100 trays in an hour.

After Care

Plug tray placement

Whether in small or larger well-established nurseries, moving the sown trays to a darkened germination room / closed chamber that is held at a mild temperature is necessary. If there is no space for a separate germination room for smaller nurseries, one of the corners can be closed off and used for the purpose. This move is necessary to keep enough moisture for germination by reducing losses in evaporation. It is usually not recommended to stack several trays on top of each other, as the top ones compress the trays underneath. When the germinating trays need to be stacked due to space limitations, the use of a ziz-zag system is advised where the upper tray does not directly reach the lower tray material. Nursery tables are used on commercial sale, the sown trays can be mounted directly on the tables.

Watering

Water quality is important, and can greatly affect seedlings' health and growth. The most common problems with the water quality are from salts, especially when using bore water.

Electrical Conductivity (EC) and pH are used to measure water quality. The ideal ranges for these are: EC should be below 1 mS / cm. Growth of seedlings can be badly affected if the water EC is small. Water pH should be 6.5–8.4. Higher pH levels are associated with higher salt levels which damage seedlings. Both nurseries need to have an EC meter and pH meter installed.

Also prefer a lower EC (less than 1 mS/cm). No correction is appropriate. If the EC is too high, the problem can be reduced through several options:

» A rainwater harvesting system can be designed to combine groundwater with rainwater to reduce the EC.

» Bactericidal water-softeners (such as potassium chloride) can be used to minimize water EC.

» Good drainage in the seedling trays will prevent salts from developing, even if the EC is not ideal at first.

Suggested time and frequency of irrigation

This can be irrigated in the morning. If seedlings are irrigated at night, droplets of water may remain on the leaves and can lead to fungal infections. If the medium is porous, then less water is required. Watering two to three times daily, depending on the environmental conditions. Irrigation during hot weather can be restricted to once daily. When you see drooping seedlings, immediately carry water. Reduce irrigation last week to help the seedlings harden before transplantation.

Nutrient supply

Fertilizers are not applied to the growth medium but the developing seedling is supplied with nutrients by regular fertilization in the artificial medium. That can be implemented in many different ways:

Basic fertilizer application

Regular application of 19:19:19 containing micronutrients (preferably Grade-4) @ 5 g per 10 liters of water, starting from cotyledon to 16 days, gradually increasing the dosage from 16 to 24 days per 3-4 days, and then stopping for hardening.

Table 6.1: Fertilizer dosage

DAS	Fertilizer dosage (g in 10 litres of water 19:19:19+ micronutrients)
8-16	5
16-19	10
19-22	15
22-24	20

Advantages of high tech nursery raising

1. Sowing can be elevated under extreme climatic conditions where otherwise it is not possible.

2. Should grow good seedlings.

3. Better grow of the root.

4. There is no risk of seedling being infected with soil-borne fungus or virus, as the nursery is grown in soilless sterilized paper.

5. Drastic decrease in seedlings transplantation mortality compared with conventional nursery elevations.

6. Planting in the early / offseason is achieved by increasing these nursery.

7. Easy to transport.

8. Weed free.

9. The seed rate in this technology can be reduced to 30-40 per cent compared to the open field.

10. Saves fertilizer and plug-tray nursery water.

Chapter - 7

Problems/ Constraints of Greenhouse Cultivation and Future Strategies

Though, India's Greenhouse production system is very ancient, but its infancy is technology. There has been little effort to harness the full potential of the greenhouse production system. Some major issues / restrictions that restrict the use of greenhouse technology are:

Major Issues

1. The basic structural and operating costs of climate-controlled greenhouses are very high which Indian farmers cannot afford without government support.

2. Uninterrupted and regular power supply is needed to operate greenhouse cooling and heating systems, which is lacking nearly everywhere in India.

3. Also little work on standardizing the structural designs for the country's various agro-climatic regions.

4. Required quality of cladding material is not readily available. In addition, replacement with new generation of biodegradable polymers is not available to address environmental issues due to the use of plastics.

5. There is a lack of appropriate instrumentation for monitoring of the greenhouse environment.

6. Non-availability of instruments and implements to facilitate greenhouse crop production operations.

7. Non-availability for greenhouse cultivation of indigenously bred / developed varieties/ hybrids.

8. Very little work on the development of greenhouse crop POPs, INM, and IPDM modules.

9. There is a shortage in the country of unique greenhouse development programs.

10. Protected agriculture increases productivity of the soil, water and fertilizer along with decreased use of pesticides. But the initial costs associated with these treatments are very high. Although it is linked to several government schemes in India and has 50-90 percent of subsidies depending on state and central government intervention, status / socio-economic status and policy.

11. Uninterrupted and regular power supply is needed to operate greenhouse cooling and heating system.

12. Required quality reclining content is not readily available.

13. Non-availability of instruments and implements to facilitate greenhouse crop production operations.

14. There is a lack of major greenhouse crop production research program at the region. In the absence of climate controlled greenhouses, specific technologies for the region are needed.

15. Lack of or scanty breeding research programs to grow suitable crop varieties / hybrids for greenhouse cultivation. Exotic seeds are very expensive and out of reach of Indian cultivators. Their daily supply is uncertain.

16. Pollination problem in greenhouse crops

17. Increasing threat of bio-stresses, especially root-knot nematodes and *Fusarium* wilt in greenhouse cultivation.

Future Strategies

1. Establishment of National Protected Cultivation Research Centre.

2. Creation of suitable cultivable varieties / hybrids.

3. Identify new and future crop varieties for greenhouse.

4. Development of greenhouse designs which are cost effective and location-specific.

5. Photovoltaic greenhouse (PVG) systems are planned and simulated for Indian conditions.

6. Creation of full POP for greenhouse crops in the country's various agro-climate regions.

7. Identification of unique, techno-economic and feasible crop sequences for the region.

8. Develop successful system INM and IPDM.

9. Production of the Good Agricultural Practices (GAP) Module

10. Design and development of greenhouse materials, instruments and equipment

11. Because of the major concern due to plastic use, the production and use of biodegradable polymers from the new generation is important.

12. Need to be working on agriculture sustainability under protected structures.

13. Standardization of new age technology such as hydroponics, aeroponics, technique of nutrient film, agro-voltaic systems, vertical farms, etc.

14. Developing cost-effective agro-techniques for growing different crops and reducing energy costs if any of maintaining the greenhouse climate.

15. Build competent and trained manufacturers of polyhouse goods.

16. Under the Ministry of Agriculture, a mission is required on protected farming in the region, and farmers' welfare.

17. Computerized control systems are needed to maximize returns: these include time base / volume base / sensor-based irrigation system, opening and closing of ventilators and side wall rolling curtains, retractable roof, CO_2 generator, climate control, temperature, humidity, heat radiation, EC control, pH, ppm level of irrigation water elements, etc., as required by the plant.

18. Managing pollination inside protected structures in vegetable crops.

19. Aeroponic systems in the production of potato seeds.

20. Work into new edible class "Microgreens"

Chapter - 8

Growing Media, Soil Pasteurization, Cost Estimation and Economic Analysis

GROWING MEDIA

Under the greenhouse production method there are two types of growing media used:

(1) Soil-based growing media

(2) Soil-less media

Soil based growing media: For crop production, most greenhouses utilize soil-based growing media. That is the easiest way to start producing greenhouse. Traditionally, the soil medium is made up similarly by volume basis of loan-filed soil, coarse sand and well-decomposed organic matter and the media is typically balanced to a pH of 6.0-6.5. Red laterite soils are highly suitable for soil-based greenhouse cultivation, as pH and EC range falls within the comfort zone of growing plants in such soils. The advantage of having soil pH in that range is easy plant nutrient availability.

The proportion of organic matter should be higher if sandy soils are to be used as a growing medium, whereas the proportion of sand should be higher for clay soil. Ideally, the growing polyhouse media should consist of 70% red laterite soil in proportion: 20% FYM and 10% rice husk / sand. Often sections of well-decomposed organic matter can be substituted with coarse organic material such as coconut husk, coarse bark, peat moss and so on, which helps increase media aeration. Such an improvement can be done to create more balance between the required characteristics of growing media such as water holding capacity, aeration, steady supply of nutrients etc.

Maintenance of soil-based media

Decomposition results in loss of organic matter from the media, so customary regular application of organic matter should be done. It can be done on a routine basis each year when the growing media is pasteurized. Depending on the nutrient and drainage status, organic matter at a rate of 5 to 10 percent of the volume of growing media is good.

However, soil-based media is associated with some disadvantages and these are:

1. Soil-based media is a carrier of many diseases, insects and weeds, and therefore requires regular media pasteurization.

2. Inappropriate soil-media combination causes aeration issues, drainage etc.

Soil-less growing media: There are various soilless media used to grow greenhouse crops and they are as follows:

a. **Peat:** This consists of the remnants of forest, marsh bog, or swamp vegetation, preserved in partly decomposed condition under water. Depending on the vegetation from which it originates, the composition of different deposits varies widely depending on the state of decomposition, mineral nutrient and acidity. It is a very stable source of organic material which holds much water and air. It does not rapidly decompose. Peat is very acidic (pH 3.5 – 4.0) and calcareous is applied to change the pH. Younger, light-colored peat offers air space better than older, darker peat. Due to its wide availability and relatively low cost, peat is the most commonly used soilless medium.

b. **Perlite:** Pertile is a grey-white silica substance of volcanic origin produced by the eruption of lava. The crude ore id crushed, screened and heated in furnaces where the tiny amount of humidity in the particles changed to steam, the particle exploded to thin, sponge like kernels. It is very light and only weighs 6 to 8 Ibs / cubic feet. The high temperature of the processing gives a sterile product. Pieces of perlite create tiny air tunnels that allow free flow of water and air to the roots. Water will adhere to the perlite surface, but the perlite aggregates are not absorbed in. It is pH neutral and has insignificant capacity for exchanging cations. While costs are moderate but perlite is an effective change to growing media.

c. **Vermiculite:** This is a micaceous mineral, expanding significantly when heated. Chemically, it is a silicate hydrated from magnesium-aluminum-iron. It is very lightweight (6 to 10 Ibs / cu. ft.), reaction neutral and water insoluble. It has a nearly natural pH. Due to its high nutrient and water retention, good aeration and low bulk density, vermiculite is very desirable component of soilless media. It is not commonly associated with soil.

d. **Sphagnum moss:** Hydrated remains of acid bog plants such as *Sphagnum pappilosum* and *Sohagnum palustre* are commercial sphagnum moss. It is relatively sterile and light in weight with a high ability to carry water. It consists of water holding cells, so that they can absorb 10 to 20 times their water weight. It also consists small amount of minerals.

e. **Leaf mold:** Maple, oak, sycamore and elm are suitable for leaf molding amongst the type of leaf. After leaves have been composted to 12 to 18 months leaf mold is ready for use. It can contain nematodes, weed seeds and noxious insects and diseases, and should therefore be sterile before use.

f. **Bark:** Bark is partially composted material of plant origin and is screened. When bark is removed from logs, it includes varying amounts of cambium and young wood. These materials decompose faster than bark and accentuate the problem of nitrogen bonding. To save shipping costs, it is beneficial to get bark from local sources.

g. **Rockwool:** Rockwool is formed by burning a mixture of coke, basalt, chalk and probably iron slog. The actual composition is around 60% basalt, 20% coke and 20% limestone. It is designed to recommend higher density to provide the need for plants to retain air and water. It is used for propagation in cubes and finishing of crops in slabs. Insulation and acoustic-grade rock wool are ideal for growing plants. The granular form has very high properties in terms of water and aeration. This is not buffered, though mildly alkaline, and has negligible CEC. This does not contribute to any extent, or carry nutrients.

h. **Polystyrene form:** It helps to bring better aeration and lighter weight to the root substrate. It is a white synthetic product which contains numerous air-filled, closed cells. It is highly lightweight, does not absorb water and does not have any appreciable CEC. It is inert, and thus does not affect pH levels of the root substrate. Polystyrene is available in beads or in flakes. Beads 3 to 10 mm in diameter and flakes 3 to 10 mm in diameter are satisfactory for pot plant substrate. This is also well resistant to decomposition.

i. **Sand:** Sand is a basic component of soil, the particle size varies from 0.05 mm to 2.0 mm in diameter. Fine sand (0.05 mm -0.25 mm) does little to boost a rising media's physical properties and can result in decreased drainage and aeration. Medium and coarse sand particles are the ones which provide optimum media texture adjustment. Although generally sand is the least expensive of all inorganic modifications. It is heaviest which can lead to prohibitive transport costs. Sand is a valuable modification to both potting and propagation media.

j. **Rice hulls:** Those are rice milling industry by-products. Rice hulls are extremely light in weight and very effective for drainage improvements. The particle size and resistance of rice hulls to decomposition, are very close. Nevertheless, N depletion in media adjusted with rice hulls does not pose much serious issue.

k. **Calcined clays:** Calcined clays are formed by heating the minerals of montmorrillonitic clay to about 690 ^{0}C. The pottery that is formed like particles is six times as heavy as perlite. They have relatively high exchange of cations, as well as capacity to hold water. The amendment is very long-lasting and useful. This inorganic soil alteration is usually used to increase the amount of large pores, decrease the capacity to retain water and boost the drainage and aeration.

l. **Bagasse:** Bagasse is a by-product of sugar industry waste. To produce material, it may be shredded and/or composted, which may increase the aeration and drainage properties of container media. Rapid microbial activity results after the incorporation of bagasse into the media, due to its high sugar content. This decreases bagasse's durability and longevity, and affects nitrogen levels. Although bagasse is readily available at low cost (usually transport), it is limited in use.

m. **Coco-peat:** It is a coconut husk by-product. Once the husk has grinded, coco-peat is ready. Currently, it is used commercially in the polyhouse pot growing of vegetables for days. It swells 15-18 times more than its original weight after soaking in water. It is more appropriate, cheaper and in ample quality. This comes in loose shape as well as packed bricks. Coco-peat is best considered for providing media aeration, drainage and life.

n. **Sawdust and shavings:** These are Lumber Mills by-products. Such materials decompose at a faster rate than bark and a greater amount of nitrogen is tied up in the root substrate due to its larger C : N ratio (1000:1). For one month, saw dust is composted with additional nitrogen to increase the nitrogen content. It continues to decompose in pots or in greenhouse benches during use. When thoroughly composted, it is near neutral in pH.

Advantages of soil-less media

1. Growers that have no field of their own are also able to grow other crops of vegetables.

2. To grow vegetables in greenhouses, kitchen gardens, houses and areas where suitable soil areas are not available, it has broad adaptability.

3. Soil-less culture media is free of nematodes and pathogens.

4. It does have decent ventilation and drainage.

5. Control of nutrients: a complete analysis of soilless crop media is possible which saves fertilizer usage.

6. Labor economy: Less time for handling, sterilization, weeding, application of fertilizers and watering.

7. The light weight of soilless substrates decreases shipping costs.

Soil Pasteurization

Under greenhouse, soil-based media may contain harmful disease causing organisms, nematodes, insects and weed seeds. Soil pasteurization or sterilization for soil-borne pathogens control should therefore be an essential component in greenhouse soil-based media. Disinfection approaches such as solarization, steam, chemical and biological control can be used efficiently.

Solarization

During warmer months of the year the soil or growing medium can be disinfected by covering the soil with transparent plastic. This will dramatically increase soil heat and kill many pathogens and insects born into the soil. It is based on trapping solar irradiation by covering the wet soil tightly, usually with transparent polyethylene or other sheets of plastic. This results in a substantial elevation of soil temperatures (10-15 °C above normal, depending on the depth of the soil) to the point that most pathogens are susceptible to heat when applied for 4-6 weeks and controlled either directly by the heat or by the chemical and biological processes produced in the heated soil.

Steam pasteurization

Water steam used since the beginning of the 20th century for soil fumigation, records of which were observed in Germany during 1888 (Baker, 1962). A wide variety of steam machines have been built since that time to sterilize greenhouse soils. The effect of water vapor as a phytoiatric mean is obtained by heating the soil to levels causing coagulation of proteins or inactivation of enzymes. The time of application exposure and the buildup of temperature in the soil / substratum are the key parameters for achieving effective results. These parameters are influenced by a variety of factors related to soil characteristics (e.g. texture, organic matter quality, water quality etc.) and application technological aspects (e.g., steam delivery system). However, high temperature steaming at 85 °C to 100 °C in the early days killed too many beneficial soil species (i.e. mycorrhizal fungi, nitrogen-fixing bacteria) along with the pathogens and led to the development of harmful phytotoxic compounds for crops. Therefore, an effective treatment currently must achieve a soil temperature of at least 70 °C for 30 minutes. Within a few minutes even a less extreme approach was suggested with a temperature of 50-60 °C. Therefore, this technology should first be loosened before pasteurization to make it effective growing medium. This will help the movement of steam through the pores and quickly transmit heat inside the medium. The root medium should not be dry, and if dry, the rate of pasteurization is increased by adding water. The excess watering may slow down the speed of pasteurization, therefore moistening of root medium a week or two prior to pasteurization is the best procedure, which breaks the dormancy of many unmanageable weed seeds and then pasteurization destroys them easily to strengthen the adoption of this technology Govt. of India is also providing 50% subsidy under a Mission for Integrated Development of Horticulture.

Fumigation or chemical pasteurization

Greenhouse growing medium may contain harmful disease causing organisms, nematodes, insects and weed seeds, so it should be decontaminated by heat treatment or by treatment with volatile chemicals such as metham sodium, formaldehyde, dazomet, basamide, chloropicrin, etc.

Table 8.1: The protocol followed by each chemical for soil-pasteurization

Chemical	Dose	Recommendation
Formaldehyde	75 litre per 1000 m^2	Irrigate the soil to the field capacity. Drenching soil with Formaldehyde (37-41%) at 1:10-Chemical: water ratio. Covering soil with black polyethylene sheet and keeping it for a week. Removal of covering after 4-5 days and irrigation of soil to drain excess of fumes, if any.
Metham sodium	30 ml/m^2	Cover the soil after application of chemical for 4-5 days with black polyethylene sheet and aerate for at least 8-10 days.
Dazomet	40 g/m^2	Pre-plant application 30 days before planting.
Chloropicrin (Tear Gas)	3-5 ml/cu. ft. of medium	Cover for 1-3 days with gas proof cover after sprinkling with water. Aerate for 14 days or until no odour is detected before using.
Basamid	8.0 g/cu.ft. of medium	Cover for 7 days with gas proof cover and aerate for at least a week before use.

Chloropicrin or tear gas is often used as a fumigant, but its low root media penetration into the plant tissue is a drawback.

Physical propagation facilities such as spread space, bins, tables, knives, working floor, benches etc. can be disinfected using one part of formalin in fifty water parts or one part of sodium hypochlorite in nine water parts. Regularly sprayed an insecticide such as dichlorvos will take care of the insects present, if any. Until going into the greenhouse with a prescribed seed treatment chemical for seeds and a fungicide-insecticide combination for cuttings and plugs respectively, care should be taken to disinfect the seed or the planting materials. Disinfectant solution such as trisodium phosphate or potassium permanganate at the greenhouse entrance will help rid the pathogens of the greenhouse entrance workers.

Safety directions while using chemical sterilants

» Harmful, if swallowed or inhaled. Toxic if inhaled. Will hurt your eyes, your nose, your throat and your skin. Repeated exposure can cause allergic troubles. Alcohol interacts-stop alcohol on day(s) of use. Do not spray mist or inhale vapour. Second, the fumes trigger smart, then eye watering. This should be viewed as a warning sign. The liquid can lead to burns. Use and store in well ventilated environments.

» Use chemical-resistant neck and wrist pants, washable hat, elbow-length chemical-resistant gloves, chemical-resistant boots, and full face-piece organic vapour respirator when opening the package and preparing the product for use.

» Workers previously experiencing irritation of the skin or respiratory tract from chemical exposure(s) should not work with such products.

» Do not allow the mixture to stand up after combining with water as toxic gases are emitted on stand.

» Workers manually sealing should wear applicator-specific personal protective cloth.

Augmentation of soil with bio-agents

Increase applies to all types of biological control in which natural enemies are added regularly and typically allow the released agents to be commercially produced. The preparation of beds by building up a rich flora of biological control agents such as *Trichoderma* spp., *Pseudomonas fluorescens*, *Paecilomyces lilacinus* etc. is important to the management of soil-borne pathogens, especially nematodes. Additionally the following precautions should be followed:

» Avoid / repair faulty greenhouse structures that aid insect-pest entry.

» Always use net screens which are insect-proof.

» Greenhouses should preferably have double entry gates in L-shaped form along with workers / visitors to minimize the risk of pest entry.

FYM increases can be achieved by mixing *Trichoderma harzianum* cultivation at 1 kg/500 kg and/or *Pseudomonas fluorescens* or *Paecilomyces lilacinus* at 50 g / m² 15 days before sowing.

Cost Estimation and Economic Analysis

Regardless of the type, protected agricultural systems are extremely expensive. The equipment and production cost may be more than compensated by the significantly higher productivity of protected agricultural systems as compared with open field agriculture. The cost and returns of protected agriculture vary greatly, depending on the system used, the location and the crop grown. By design, all protected agricultural systems of cropping are intensive in use of land, labour and capital. Greenhouse agriculture is the most intensive system of all. The intensity of land use is greatly dependent upon the system of protected agriculture. Year-round greenhouse crop production is therefore much more intensive than seasonal use of mulches and row covers. Coinciding with intensity are yields, which are normally far greater per ha from year round than from seasonal systems. The normal benefit of higher yields of Controlled Environment Agriculture (CEA) over the open field agriculture depends on the system used and the region of production.

Capital requirements

The capital requirements differ greatly among the various systems of protected agriculture. Mulching is least expensive while greenhouses require the most capital per unit of land. Total cost involved in the production is the sum of fixed cost and operating cost as shown (Fig 8.1). The fixed capital costs include land, fixed and mobile equipment, and structures like grading, packing and office. Fixed costs also include taxes and maintenance. The fixed capital costs for greenhouses clearly exceed those of other systems of protected agriculture, but vary in expense according to type of structure, and environmental control and growing systems. Operating costs include labour, fuel, utilities, farm chemicals and packaging materials. The operating or variable costs and fixed costs are annual expenditures and these can be substantial. Annual costs may correlate to some extent with capital investment. The flow diagram of capital requirements of production is shown in figure.

In estimating the capital requirements, the farmer must include the cost of the entire system as well as the mulch. While greenhouse production systems may be far more expensive than open field systems of equal land area, open field systems of protected agriculture are normally more expensive in field area than in greenhouse production. Greenhouses are expensive, especially if the environment is controlled by the use of heaters, fan and pad cooling systems and computer controls.

Economics of production

Production economics considers the various components of fixed and variable costs, compares them with the income and evaluates the net return, on unit area basis. On an average basis, wages account for approximately 85% of the total variable cost. Wages are the greatest expenditure in greenhouse production, followed by amortization costs and then energy costs, and energy expenditure, when heating is necessary. About two-fifths of the expenses are fixed costs and about three-fifths are variable costs. Depreciation and interest on investment accounts for most of the fixed costs.

Conditions influencing returns

A number of variables which may not show up in the yearly financial balance sheet influence the returns to greenhouse operators, such as economics of scale, physical facilities, cropping patterns and government incentives. The size of any system of protected agriculture will depend on the market objectives of the farmer. Most protected agricultural endeavors are family operated. Often the products are retailed directly to the consumer through a road side market at the farm site. In the developed world, greenhouse operations tend to be a size that can be operated by one family (0.4 to 0.8 ha). A unit of 0.4 ha can be operated by two to three labourers, with additional help at periods of peak activity. The labour wages can usually be provided by the owner and his family. Moreover, the owner will pay close attention to management, which is the most important factor. Labour costs may rise significantly if it is necessary to recruit labour from outside the family. Green house owners who hire a highly qualified manager may have to operate a larger greenhouse than family size greenhouses in order to offset the additional salary paid.

The green house system economy can be improved with increased size when:

1. There is a unique opportunity to mechanize certain operations.

2. Labour can be more efficiently utilized.

3. Low cost capital is available.

4. There are economics in the purchase of packaging materials and in marketing.

5. Some special management skills are available.

The physical facilities and location of the green house influence the economics. Another variable that influence the profits from the green house is intensity of production, which is determined by the structures with complete environmental control system facilities year round production and early harvest, thus enabling the grower to realize higher profits. Year round production offers year round employment to the laborers. It is found that the environmentally controlled greenhouse produced only one third more revenue than high tunnel structure. With the improved transportation facilities, the new areas of production in combination with the following factors contribute to the lower costs.

1. High sun light intensity undiminished by air pollution.

2. Mild winter temperatures.

3. Infrequent violent weather conditions.

4. Low humidity during the summer for cooling.

5. Availability of water with low salinity levels.

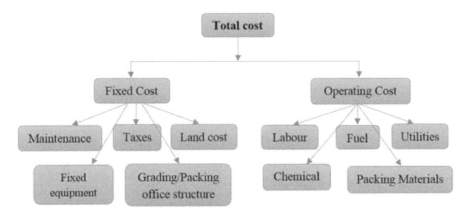

Fig. 8.1: Flow diagram of capital requirements of production

Cropping pattern will have bearing on the greenhouse structure. A high tunnel structure or any structure not fitted with environmental controlled equipment for heating and cooling will be used only on a seasonal basis. It is common to switch over from green house vegetable production to flower production, especially in structures with more elaborate environmental control systems. Growers throughout the world are currently experimenting with alternative crops, such as herbs. As eating habits change, with times and as the consumers are becoming increasingly conscious of diet and the nutritional value of fruits and vegetables, growers must continually look for alternative to existing cropping patterns.

Government policies also influence the financial returns from the crops. The component of protected cultivation is being strengthened under **Mission for Integrated Development of Horticulture (MIDH)** by Government of India through 50% subsidy to the farmers. Incentives in terms of subsidy to the tune of 65 and 75% are further disseminated by Government of Gujarat State (India) to encourage the farmers for adopting protected cultivation by adding its share of 15 and 25% in Union Government subsidy depending upon socio-economic status of the farmers.

Table 8.2: Cost norms and pattern of assistance under Mission for Integrated Development of Horticulture (MIDH) during XII Plan for Protected Cultivation

Sr. No.	Item	Cost Norms	Pattern of Assistance
Green House structure			
1	Fan & Pad system	Rs. 1650/m² (up to area 500 m²)	50% of cost for a maximum area of 4000 m² per beneficiary.
		Rs. 1465/ m² (>500 m² up to 1008 m²)	
		Rs. 1420/ m² (>1008 m² up to 2080 m²)	
		Rs. 1400/ m² (>2080 m² upto 4000 m²)	
	* Above rates will be 15% higher for hilly areas.		
2	Naturally ventilated system		
	i) Tubular structure	Rs.1060/ m² (up to area 500 m²)	50% of cost limited 4000 sq. m. per beneficiary.
		Rs. 935/ m² (>500 m²upto 1008 m²)	
		Rs. 935/ m² (>500 m² up to 1008 m²)	
		Rs. 890/ m² (>1008 m² upto 2080 m²)	
		Rs. 844/m² (>2080 m² upto 4000 m²)	
	* Above rates will be 15% higher for hilly areas.		
	ii) Wooden structure	Rs. 540/ m² Rs. 621/ m² for hilly areas	50% of the cost limited to 20 units per beneficiary (each unit not to exceed 200 m²).
	iii) Bamboo structure	Rs. 450/ m² Rs. 518/ m² for hilly areas	50% of the cost limited to 20 units per beneficiary (each unit should not exceed 200 m²).

3	Shade Net House		
	(a) Tubular structure	Rs. 710/ m² and Rs. 816/ m² for hilly areas	50% of cost limited to 4000 m² per beneficiary.
	(b) Wooden structure	Rs. 492/ m² Rs. 566/ m² for hilly areas	50% of cost limited to 20 units per beneficiary (each unit not to exceed 200 m²).
	(c) Bamboo structure	Rs.360/ m²and Rs.414/ m² for hilly areas	50% of cost limited to 20 units per beneficiary (each unit not to exceed 200 m²).
4.	Plastic Tunnels	Rs. 60/ m² Rs.75/ m² for hilly areas.	50% of cost limited 1000 m² per beneficiary.
5.	Walk in tunnels	Rs. 600/m²	50% of the cost limited to 5 units per beneficiary (each unit not to exceed 800 m²).
6.	Cost of planting material & cultivation of high value vegetables grown in poly house	Rs.140/ m²	50% of cost limited to 4000 m² per beneficiary.
7.	Cost of planting material & cultivation of Orchid & Anthurium under poly house/ shade net house.	Rs. 700/ m²	50% of cost limited to 4000 m² per beneficiary.
8.	Cost of planting material & cultivation of Carnation & Gerbera under poly house/shade net house.	Rs. 610/ m²	50% of cost limited to 4000 m² per beneficiary.

9.	Cost of planting material & cultivation of Rose and lilium under poly house/ shade net house	Rs. 426/ m²	50% of cost limited to 4000 m² per beneficiary
10.	Plastic Mulching	Rs. 32,000/ha and Rs. 36,800/ha for hilly areas	50% of the total cost limited to 2 ha per beneficiary.

Chapter - 9

Hydroponics for
Hi-tech Horticulture

SOIL-LESS CULTIVATION

Soilless culture can be defined as "any plant growing method without the use of soil as a rooting medium in which the inorganic nutrients absorbed by the roots are supplied through irrigation water." The fertilizers containing the nutrients to be supplied to the crop are dissolved in the irrigation water at the correct concentration and the resulting solution is referred to as "nutrient solution."

Over the last three decades, the rapid expansion of hydroponic systems worldwide can be attributed to their independence from the soil and its associated problems, i.e. the presence of soil-borne pathogens at the beginning of the crop and the decline in soil structure and fertility due to continuous cultivation with the same or relative crop. Cultivation of soilless appears to be the safest and most effective alternative to soil disinfection through chemical agents. In protected cultivation, therefore, it is becoming increasingly important – not only in modern, fully equipped glasshouses, but also in simple greenhouse constructions designed to optimize favorable climate conditions. **Hydroponics** (or soilless culture) is a broad term that includes all techniques for growing plants in solid media other than soil (substrate culture) or in aerated nutrient solution (water culture).

Advantages of hydroponics

1. Soil-borne pathogens are absent.

2. Soil disinfection is a safe alternative.

3. Possibility of growing greenhouse crops and achieving high yields and good quality, including in saline or sodium soils or in poorly developed non-arable soils (accounting for most of the world's cultivable land).

4. Precise nutrition control, particularly in crops grown on inert substrates or in pure nutrient solution (also in soilless crops grown in chemically active growing media, plant nutrition can be better controlled than in soil grown crops, due to the limited media volume per plant and homogeneous media constitution).

5. Eviting soil tillage and preparation, thereby increasing crop length and total greenhouse yield.

6. Early yield increase in crops planted during the cold season, due to higher root zone temperatures during the day.

7. Respect for environmental policies (e.g. reduction of fertilizer application and restriction or removal of greenhouse nutrient leaching) – thus, the installation of closed hydroponic systems in greenhouses is mandatory in many countries, especially in environmentally sensitive areas or those with limited water supplies.

Despite the significant advantages of commercial soilless culture, in some cases there are disadvantages which limit its expansion:

1. Strong costs to install.

2. Needs for professional skills.

Table 9.1: Classification of soilless culture systems

Soilless culture	
Water culture or hydroponic	**Substrate culture**
» Deep water culture	» Gravel culture
» Float hydroponic	» Sand culture
» Nutrient film technique	» Bag culture
» Deep flow technique	» Container culture
» Aeroponics	Trough culture

Hydroponic method is also graded as an open system when the nutrient solution is supplied to the plants and not reused, and when the surplus is stored, replenished and recycled, it is closed system. In these agricultural systems, since controlling the aerial and root environment is a major concern, development takes place within enclosures built to regulate air and root temperatures, light water, plant nutrition and adverse climate.

Functions of the general hydroponic system

1. It will offer power of temperature.

2. Evaporation helps to reduce water loss.

3. Reduce disease and infestation with the pests.

4. This would protect crops against environmental conditions, such as wind and rain.

Hydroponics Techniques

1. Deep water culture (DWC)

DWC, created in 1929 by Professor W.F. Gericke of the University of California, was the first hydroponic method proposed for commercial purposes. It consists of a bucket filled with nutrient solution, covered with a net and a cloth on which a thin layer of sand (1 cm) is put to protect the plants; in the nutrient solution the roots are suspended. Alternatively, a lid may cover the bucket, and the plants, contained in net pots, suspended from the center of the cover. The system's main downside is the hypoxic conditions that exist at root level due to the small exchange area between air and water relative to the solution volume, and the low oxygen diffusion coefficient in water. This restriction was overcome by air pumps that oxygenate the nutrient solution or by recirculating deep water culture systems (RDWC) that use a reservoir to supply several buckets with a nutrient solution. In RDWC it is broken up and aerated with the use of spray nozzles as the water is reintroduced to the reservoir.

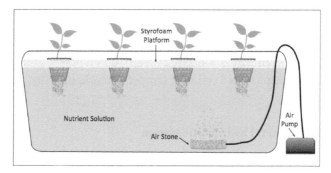

Fig. 9.1: Deep water culture

Source: https://www.kompactfarms.com/

2. Float hydroponics

Plants are grown on trays filled with nutrient solution, floating in tanks. Today the technique is primarily used for growing fresh-cut leafy vegetables (spinach, chicory, spinach, etc.) and aromatics (basil, mint, etc.).

Due to the low setup and management costs and the little automation required to monitor and adjust the nutrient solution, the system seems particularly interesting. The large volume of nutrient solution buffers the temperature and reduces the adjustment frequency and the solution's reintegration. The concentration of O_2 in nutrient solution will vary between 5 and 6 mg per litre during the growing season.

A single-tank or multi-tank system may be used at the greenhouse. The former, which covers almost the entire span, reduces the incidence of barren areas and allows the automation of certain operations, such as the placement and removal of floating trays; the latter consists of several tanks of approximately 4 m² (2 µ2 m) and reduces the risk of operational errors and diseases.

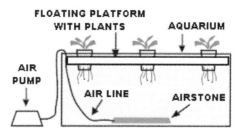

Fig. 9.2: Float hydroponics

Source: http://www.simplyhydro.com/

3. Nutrient film technique (NFT)

NFT is a hydroponic technique by which a very thin layer of nutrient solution (film) flows through watertight channels (also known as gullies, troughs or gutters), where the plant's bare roots lie. At the elevated end, the nutrient solution is applied to keep the roots completely wet, so that the solution flows down through the channels. The slope may be created by the floor itself, or the channels may be kept by benches or racks and provide the elevation necessary. The thin water stream (1–2 mm deep) ensures adequate oxygenation of the roots, because the thick root mat that forms at the bottom of the channel continually exposes its upper surface to the air. The solution is drained onto a large catchment pipe at the lower end of the channels, which leads the solution back to the cistern to be recirculated.

Channels generally consist of different types of plastic material, such as polyethylene liner, polyvinylchloride (PVC) and polypropylene, with a triangle-shaped section rectangleore. The channel base must be smooth and not angled, in order to maintain a shallow liquid current.

In general, the length does not exceed 12–16 metres. A modified method called super nutrient film technique (SNFT) has been developed to solve these problems: the nutrient solution is dispersed through nozzles positioned along the path, ensuring sufficient availability of both nutrients and oxygen near the roots.

Nutrient solution delivery can be continuous or intermittent in a 24-hour cycle (alternating watering and dry cycles to boost root system oxygenation). Another option – a compromise between these two methods – is continuous nutrient solution recirculation during daylight hours (dawn to dusk) and automatic night shutdown. Plants intended for use in NFT systems are lifted in small pots or plugs or in cubes of rockwool and put in the channels when a significant root system has been created.

NFT's main advantages over other systems are the absence of substrate and the reduced volume of nutrient solution required, resulting in significant water and fertilizer savings, and reduced environmental impact and substrate disposal-related costs. On the other hand, the nutrient solution is subjected to major changes in temperature along the canal and during growing seasons due to the low water volume. In addition, NFT has very little buffering against water and nutrient supply interruptions, so there is a significant risk of the spread of root-borne diseases. Technically most crops could be grown in an NFT system, but it works best for short-term crops (30–50 days), like lettuce, because plants are ready to harvest before their root mass fills the channel.

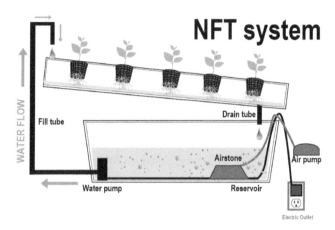

Fig. 9.3: Nutrient film technique (NFT)
Source: http://hydroponicsfarm.weebly.com/

4. Deep flow technique (DFT)

DFT is another form in which the roots are constantly exposed to moving nutrients and water. Although the water stream with NFT is as thin as possible, the continually flowing nutrient solution at DFT has a depth of 50–150 mm. The large volume of water simplifies the regulation of the nutrient solution and buffers the temperature, thereby making the device ideal for regions where nutrient solution temperature fluctuation can be a problem. In a DFT system the channel width is usually about 1 m. Plants are grown on trays of polystyrene which float on the water or rest on the sidewalls of the channel.

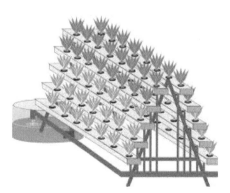

Fig. 9.4: Deep flow technique (DFT)
Source: http://www.yoonhidroponik.web.id/

When the first "agricultural revolution" began when human beings started growing crops in soil, the second could be just when we started digging them out of the ground. Of course, we are talking about hydroponics-the revolutionary, soil-free method of growing crops, vegetables, and that has sparked a new wave of excitement in the agricultural sector.

One big reason for all the interest in hydroponics: the majority of modern systems are planned not only to maximize profitability but also to minimize long-term costs. If it is by resource management or the conventional elimination of time-consuming manual labor, there is a lot for farmers to like about hydroponic technology tomorrow.

Here's a look at just a few big ways that hydroponics help growers cut down on long-term operational costs, and why it could be just time to set up your growing activity with a hydroponic system.

1. Improve water efficiency

Usually a grower who deals with plants in the conventional dirt medium would expect to lose a large portion of the water that they distribute to rain over the plants. It may not only contribute to leaching and pollution of local water sources, but it also reflects a reduction of operating income that has historically been seen as inevitable.

Because hydroponics eliminate the need to plant in the dirt or in the ground and use water itself as the main growing medium, these systems make it significantly easier for growers to reduce water runoff and increase season after season water efficiency. In a closed loop all the water in your hydroponic system persists, ensuring that just about every drop is used to the fullest degree without being lost to gravity or other thirsty plants.

2. Maximize profitable space

One of the biggest concerns we hear from farmers is that they are struggling to find the balance between "too much unused land space" and "too little room for plants to flourish." Add to that the normal unproductive room present in many conventional growing operations and it can be difficult to reach the sweet spot in terms of profit-per-square foot.

That's why hydroponic systems – with their compact spacing and highly efficient use of horizontal ground space – can provide a big boost to square-foot efficiency in your greenhouse or high tunnel structure. By removing wasted space in favor of healthy, concentrated growing areas, farmers can yield larger and better harvests using the same amount of ground space.

3. Reduced pests & diseases

Soil remains one of the principal vectors for pests, diseases, and other harmful pathogens to enter your plants in almost any conventional growing process. Whether it's because of water drainage from an impacted field elsewhere or simply because of the normal cycle of soil health, growing in-ground will almost always expose plants to the risk of disease and pests-while hydroponics, on the other hand, will help to do the opposite.

A hydroponic system is a closed system, which means that there will be no simple entry or exit points for pests, viruses, molds, or other pathogens in order to enter the plants.

As long as you and your growers maintain industry-standard cleaning and sanitation practices between plantings, your crops should be significantly easier to keep pest-free than crops grown in soil. That means severe pest prevention savings, and a significantly reduced risk of loss due to disease or contamination.

4. No weeding needed

Pests and diseases are not the only invaders trying to use your growing crops. Weeds can creep into your crop beds faster when planted in the soil than you can pull them out. That can eat up time, energy and treatment you would not have to spare otherwise.

Since hydroponic systems are soil-free and closed, weeds would have no chance of growing first in your crops. That means you can spend less time worrying about weeds and other organic invaders choking off your crops and more time ensuring that your harvest is as full and beautiful as possible.

5. Grow more profitable plants

Often times when you choose to grow in the ground that means selecting only those crops that are most likely to succeed in your area. For example, not all soil is ideal for growing grapes, while climate conditions and temperatures can make it harder to grow even simple crops in especially cold or storm-prone areas.

In comparison, hydroponics do not rely on outside conditions for proper operation. That means you can set your sights on growing even more profitable crops with the right hydroponic system than ever before, positioning yourself to increase the profit-per-square-foot and decrease waste energy when harvesting time comes.

6. Four season farming

Growers have been shackled to the changing seasons and fluctuating weather for too long when deciding when and where their crops should grow. That has traditionally meant limiting the growing season to spring and summer, or limiting what can be successfully grown in the autumn and winter seasons, in northern climates or areas likely to experience storms.

Since hydroponic systems allow even control of temperature, nutrient levels, humidity, ventilation, and other factors every day, growers choosing hydroponics do not have to restrict their winter or autumn crops to those that are likely to thrive in cold weather only. Go ahead and grow tomatoes in january-they might even come out juicier than ever before thanks to your hydro system.

7. Achieve bigger, more reliable harvests

While the opportunity to grow crops in the offseason and reduce the risk of losing crops to diseases or pests can reflect their own savings to your bottom line, increasing your harvest's size and quality can also help increase your profit-per-square-foot even more.

Hydroponics allows growers to do just that, often boosting harvest reliability and quality thanks to superior supply of nutrients and water. When each plant has direct root access to everything it needs to grow and thrive, your greenhouse is much better than you would otherwise be able to achieve in the soil.

8. Automate & save labor

One of the major drains on the budget of any grower is the sheer labor cost needed to bring a crop bed from seed or sprout to harvest. This work can include everything from weeding and combating pests to simply monitoring and adjusting environmental controls to keeping things to the optimum level. By automating your hydroponic system, you are able to drastically reduce the work required to achieve the same-or even better-results.

In a hydroponic system, anything from rising room temperature to levels of humidity and light, to nutrient and quality rates can be automated for optimal distribution even when you are miles from your greenhouse. That's because automated systems will continuously monitor your growing space for conditions change and automatically make corrections, saving you time and effort to track and respond to growing space problems.

Table 9.2: Advantages and disadvantages of hydroponics for vegetable production

Advantages	Disadvantages
Soil preparation and weeding is reduced or eliminated.	A high level of expertise is required.
Hydroponically produced vegetables can be of high quality and need little washing.	Hydroponic production is management, capital and labour intensive.
It is possible to produce very high yields of vegetables on a small area because an environment optimal for plant growth is created. All the nutrients and water that the plants need, are available at all times.	Daily attention is necessary.
One does not need good soil to grow vegetables.	Specially formulated, soluble nutrients must always be used.
Water is used efficiently.	Pests and diseases remain a big risk.
Pollution of soil with unused nutrients is greatly reduced	Finding a market can be a problem.

Table 9.3: The difference between hydroponic production and production in soil

Hydroponics	Field production
No soil is needed.	Good topsoil is required. Good soil means good drainage, compost, disease-free.
Plants are irrigated automatically. No water stress.	Plants need to be irrigated to minimise water stress
Nutrients are available at all times Only soluble fertilizers are used. Hydroponic fertilizer formulations contain a balanced nutrient content	Nutrients must be added to soil. Unless a laboratory analysis is done, too much or too little nutrients can be added.
Soil borne diseases can be eliminated	Soil borne diseases can build up in the soil.
Hydroponic production is not organic because artificial nutrients are always used and plants are usually not grown in soil.	It is possible to produce organic vegetables in soil because one can use organic fertilizers such as compost and manure.

Aeroponics

Aeroponics is the method of growing plants in the air or in the mist without the use of soil or accumulated media. The term aeroponic is derived from the Latin term 'aero' (air) and 'ponic' means labour (work). This is an alternative form of soil-less agriculture in growth-controlled conditions. Aeroponics culture varies from traditional hydroponics, aquaponics and in vitro (plant tissue culture) growing. Aeroponics is now being effectively used in South America, and efforts are being made to implement this method in several African countries as well. Various soil-less processing methods, such as aeroponics and Nutrient Film Techniques, have been introduced in modern horticulture. Earlier study has demonstrated positive success with NFT in the processing of potato tubers. However, tuber initiation was worse in stable media-free nutrient solution than in porous media (e.g. perlite or vermiculite). Inhibitions of tuberization of stolons immersed in a solution may have the effects of a lack of mechanical resistance. The use of aeroponic systems for the processing of potato seed is very recent in Europe. Until 10 years ago, the use of these innovations was limited almost anywhere in the world, and only few nations, such as China or Korea, used them for the industrial production of quality potato seeds.

Aeroponic cultivation technique is an optional method for soil-less cultivation in growth-controlled conditions such as greenhouses. This approach consists of enclosing the root structure in a dark chamber and providing the fog unit with a nutrient solution. This was commonly used in horticultural organisms, including tomatoes, cabbage, cucumber and ornamental plants such as chrysanthemum or poinsettia. Aeroponic seed production systems have been developed following increased demand for more reliable , high-quality seed production methods. Aeroponic device has been widely used in the cultivation of potato seed tubers in Korea. As a result, aeroponics or aerohydoponics have replaced conventional hydroponic systems for the development of mini tubers. Despite growing interest in land with fewer farming methods for commercial horticultural production, little knowledge is available for potatoes. Aeroponic systems are more effective in terms of water supplies than the hydroponic system. Another unique feature of aeroponics is the limited interaction between the support system and the plant, which makes it easy to expand the plant unrestrictedly. Aeroponics devices are commonly used in NASA space exploration programmes.

Aeroponics is the growing of plants with the root system suspended in a fine mist of nutrient solution applied continuously or intermittently. Plants are secured in holes on polystyrene panels using polyurethane foam: panels are placed horizontally or on a slope, and fixed over a metal frame, arranging closed containers with a square or triangular section.

Water and nutrients are supplied by spraying the plants' hanging roots with an atomized nutrient solution by means of sprayers, misters or foggers inserted in PE or PVC pipes placed in the unit. Spraying usually lasts 30-60 seconds, and their frequency varies according to species, plant growth stage, growing season and time of day (e.g. in summer, during rapid vegetative growth, a crop grown in northeast Italy may require up to 80 sprayings per day). At each nebulization, the drainage is collected at the bottom of the modules and recirculated.

Aeroponics permits a major reduction in water and fertilizer consumption and ensures adequate oxygenation of the roots. However, aeroponically grown plants may experience severe thermal stresses, especially in summer. Another disadvantage is the inability to buffer interruptions in the flow of nutrient solution (e.g. power outages). Aeroponics may be used for small-sized vegetables (e.g. lettuce and strawberries) and medicinal and aromatic plants.

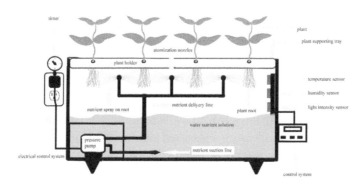

Fig. 9.5: Aeroponics system
Source: Lakhiar *et al.* (2018)

History

Techniques of soilless plants were first introduced in the 1920s by botanists who used rudimentary aeroponics to study the root structure of plants; aeroponics has since been used as a root physiology research method. Soilless culture is considered a new phenomenon, but growing plants in above ground containers have been attempted at different times over the years. The first known case of container-grown plants seems to be the wall paintings found in the Temple of Deirel Bahari. In the early 1940s, technology was mainly used as a testing instrument rather than a commercially viable crop production process. Ok, it was W. Carter, who first studied air culture in 1942, identified the technique of growing plants in water vapour to allow root testing. L.J. in 1944. Klotz was the first to find the vapour misted citrus plants in his facilitated study into the diseases of citrus and avocado roots. G.F. in 1952. Trowel grew apple trees in a spray field. Fifteen years after the Carter (1942) and Went (1957) experiments, the air-growing method of spray culture was referred to as "aeroponics." The first commercial aeroponics rig was the GTi Genesis Rooting Method, generally referred to as the Genesis Unit, in 1983. The unit was powered by a microchip and clearly connected to an electrical outlet and a water tap. Aeroponics has been extensively used in the cultivation of many ornamental and horticultural crops. Aeroponic device has been widely used in the cultivation of potato seed tubers in Korea. Aeroponics has been used in agriculture across the globe since 2006. Farran and Mingo (2006) recorded a minituber yield of 800 tubers / m² at a plant density of 60 plants / m2 over a five-month span with weekly harvests. This corresponds to a multiplication rate of 1:13. The field output of aeronically grown tubers was also found to be similar to minitubers grown from containers. At the International Potato Center (CIP) in Peru, yields of more than 100 tubers / plant have been obtained.

Importance of Aeroponics in Vegetable Crops

Aeroponics is an advancement of artificial life support for non-damaging plant support, seed germination, environmental management and drip irrigation methods that have been used by conventional farmers for decades. The biggest benefit of Aeroponics is exceptional aeration. These methods have been given particular consideration by NASA, as the fog is simpler to manage than the liquid in a zero-gravity environment. One of the biggest downside was costly.

Growing in soil is no longer a safe way to produce food for the world's 7 billion inhabitants. Increase crop yields by 45% to 75%. The effects in using the aeroponics method across their counter parts are more effective use of space. About 99 percent of the water is being used. Since pesticides and soil-compatible fertilisers are not used, the fruit and vegetables produced are pure and do not need to be washed until usage. It delivers nutrients directly to the roots of the plant, resulting in faster crop growth. Fruits and vegetables from the greenhouse are good, safe, natural, rich, fresh and tasty. Uniform growth of all crops has also been observed.

Aeroponic bio-pharming is used to produce prescription drugs within plants. The system enables the completed containment to remain inside a closed-loop network. Reports suggest that the method is ten times more effective than traditional methods, tissue culture and hydroponics, which take longer and are often more labour-intensive. The machine has the potential to save water and electricity. The Aeroponics device uses nutrient solution recirculation, so a small quantity of water is required. Comparatively, it has lower water and energy inputs per unit of increasing area. The use of aeroponics for cloning increases root growth, survival rate, growth rate and maturation time. Studies have shown that the average tuber yield under aeroponics is higher than when the same material is left to produce tubers under traditional means. These findings explicitly demonstrate that the aeroponics system can be used successfully for the spread of potatoes. The aeroponic device optimises root aeration. This is so because the plant is completely suspended in the air, allowing the plant stem and root systems access to 100% of the available oxygen in the air that encourages root growth. Such an atmosphere also allows plants 100 per cent access to carbon dioxide concentrations ranging from 450 to 780 ppm for photosynthesis, so plants in the aeroponics are developing quicker and consuming more nutrients than normal hydroponics plants. The aeroponics method improved the stomatological conductivity of the leaf, the intercellular concentration of CO_2, the net photosynthetic intensity and the photochemical efficiency of the leaf.

The Aeroponics method of propagation is one of the quickest ways of replication of seed. A single potato plant can grow more than 100 minitubers in a single row, as opposed to the traditional approach of growing approximately 8 daughter tubers in a single year, although only 5 or 6 tubers per plant are produced in a greenhouse in 90 days. Another benefit of the aeroponics method is that it is possible to track nutrients and pH. The aeroponics device provides correct plant nutrient needs for the seed, eliminating fertiliser requirements and mitigating the possibility of surplus fertiliser residues flowing into the subterranean water table.

In comparison to other techniques such as hydroponics and traditional systems, aeroponics uses better vertical space for root and tuber growth. The aeroponic atmosphere is kept free from pests and diseases as the interaction with plants is also reduced; plants grow better and faster than medium-sized plants. In addition, once the plant gets diseased, it is easily withdrawn from the support system of the plant without damaging or infecting the other plants.

As a result, many plants can grow at a higher density (plants per unit area) than in conventional modes of agriculture, such as hydroponics and soil. The method clones plants in less time and decreases the amount of work measures involved with other methods such as tissue culture techniques. In addition, the air-rooted plants are cloned and transplanted straight into the field without risk of seedlings being vulnerable to wilting and leaf loss due to transplant shock.

Table 9.3: Difference between aeroponics and hydroponics

Aeroponics	Hydroponics
Roots: Suspended in air or in an enclosed environment, Thus, the plants are able to absorb the nutrient-rich water solution and is able to remain oxygenated	Roots: Immersed in a nutrient-rich medium such as water or soil or may be supported by inert media such as gravel or perlite, Thus, the plants are able to absorb the dissolved nutrient in the medium
Solution: Sprayed onto the fine mist of mineral nutrients	Solution: Dissolved in the medium
Crop Yield: Harvest better quality and more food due to better aeration available to roots	Crop Yield: Harvest poorer quality and less food due to a limited amount of air and nutrient
Exposure to CO_2: Greater and Larger exposure	Exposure to CO_2: Smaller and Less exposure
Spread of Diseases: Reduced	Spread of Diseases: Possible
Water: Required in minimal amount	Water: Required twice the amount of water by aeroponics

Nutrients used in aeroponics system

The indoor aeroponics method uses fewer water and nutrients so the roots of the plants are sprayed at times using a precise drop size that can be used more effectively by osmosis to feed the plant. Little excess nutrient solution is wasted due to evaporation or drainage. Plant disease is reduced when the roots are kept exposed to the sunlight, preventing soaking is a stagnating moist process. Aeroponics makes it possible to grow plants without soil or substrate, to achieve maximum yields, to conserve water and fertiliser solutions and not to contaminate the ecosystem. Carbon, oxygen and hydrogen are found in the air and in the sea. Water may contain a number of elements with primary nutrients such as nitrogen, phosphorus, potassium and secondary nutrients such as calcium , magnesium and sulphur, micronutrients such as iron, zinc molybdenum, manganese, boron, copper , cobalt and chlorine. Roots utilise nutrients such as ions of water that are positively charged cations or negatively charged anions. Ammonium, NH^{4+}, and anion nitrate, NO^{3-}, both essential sources of nitrogen for plants, are examples of cation. When ions are used by plants, the pH of the solution will change, which means that it can be too positive or too negative. The optimum pH for plant growth is between 5.8 and 6.3. In aeroponic systems where water and nutrients are recycled, it is necessary to calculate acid / base or pH in order to allow plants to absorb nutrients. Aeroponic, using a mist to nourish roots, requires far less moisture resulting in simpler control of nutrient amounts and greater pH consistency.

Table 9.4: The main nutrients used in aeroponics

Nutrient	Concentration (g/L)
$N-NH_4$	0.54
$N-NO_3$	0.35
P	0.40
K	0.35
Ca	0.17
Mg	0.08
Na	0.04
Fe	0.09
Zn	0.03
B	0.03
Cu	0.04

Aeroponics growing system

The concepts of aeroponics are based on the idea of growing vegetables whose roots are not inserted in a substratum (in the case of hydroponics) or soil, but in a reservoir filled with streaming plant nutrients. In these barrels, the root can find the best position for the best oxygenation and moisture content. These conditions allow for improved assimilation of plant nutrients in a more controlled manner, resulting in faster growth of the cultivated plant.

Plant containers can be placed on top of each other and, since they are lightweight and convenient, they can be quickly moved according to agricultural needs. Numerous plants are placed in vertical columns in a greenhouse or shade house room. Nutrients are allowed to trickle down the growth columns. During the first vegetative growth, most agricultural plants require direct exposure to the sun. After that, this direct exposure is no longer relevant. Based on this observation, plant containers are regularly moved. Young plants are positioned at the highest level in the growth column. Afterwards, they are slowly lowered using a rotary mechanical system. With the rotation regularly repeated, this enables continuous development without any interruption. The Aeroponic method is a non-stop production-cycle agriculture.

Plant nutrition is given in a closed circuit. As a result, the intake is limited to only the amounts produced by the plants, making for considerable savings in water. For example: growing a kilogramme of tomatoes using conventional land takes 200 to 400 litres of water, hydroponics requires about 70 litres, aeroponics needs just about 20 litres. Since the aeroponic device is a continuous motion in the confined space, it reduces agricultural work to a collection of mechanical repetitive operating activities that are carried out on a daily and year-round basis. This allows workers to develop significant expertise within a limited span of time of a few months. In conventional agriculture, industrial production is achieved only through professional workers with several years of experience.

The aeroponic equipment is shielded within the latitude of the greenhouses or anti-hail-storm coverings. Climate monitors within the greenhouse ensure ideal growth conditions, producing good yields.

Components of Aeroponics System

Spray misters

Atomisation is accomplished by pumping water into the nozzles at high pressure. The nozzles come in various spray shapes and orifices. Larger nozzles and orifices minimise the risk of clogging, but require strength to work and high flow speeds. Selecting the nozzles that generate the appropriate droplet size would provide sufficient coverage at the expected rate and pressure.

The size of the droplet in the spray can range from sub microns to thousands of microns. This droplets are classified into various classifications. In the case of HPA, the type is thin atomization thin mist of 10 to 100 um. Using a fine mesh philtre to avoid clogging until the misting of the nozzles. Hydro atomize water and nutrient solution to 5-50 micron droplet spray size. Jet nozzles with 0.025 inch orifice working at 80 to 100 psi produce 5-50 micron droplets per second. Spray jet with 0.016 orifice running at 80 to 100 psi produces 5-25 micron droplets per second.

Droplet size

The average droplet size range for most species of plants is 20-100 microns. In this range, the smaller droplets saturate the air, preserving the humidity levels inside the growth chamber. The largest droplets, 30-100 microns, make the most contact with the base. Spray droplets of fewer than 30 microns appear to linger as fog in the air. While any droplets of more than 100 microns tend to fall out of the air before they contain any roots. So big a water droplet means less oxygen is accessible to the root organ.

High pressure water pump

High-pressure aeroponics requires a pump that can generate enough to pressurise water to generate an optimal droplet size of 20 to 50 microns. These pumps are usually diaphragm or reverse osmosis booster pumps. The pump shall generate a steady 80 PSI with the necessary nutrient supply.

pH meter

The optimum pH for plant growth is between 5.8 and 6.5. In aeroponic systems where water and nutrients are recycled, it is necessary to calculate pH in order to allow plants to absorb nutrients. Nitrogen (N) is best absorbed at pH 6.0, while phosphorus (P) and potassium (K) are best absorbed at pH 6.25 and above. The pH scale is a way to calculate acid or alkaline in a nutrient solution. The official pH definition is: a unit of measurement that defines the degree of acidity or alkalinity of a liquid solution. It is measured on a scale from 0 to 14. Acids vary from 0 to 7, with lower numbers being stronger acids. Alkaline is in the range of 7 to 14, with higher numbers being a better basis. When you test your pH, you want to blend the nutrients with the water absolutely first to ensure true reading. If reading is not at the right stage, you need to change it using pH-adjusters named "pH up" and "pH down," depending on the weather, the reading is too high or too low. If it is too high, use the pH down, and if it is too low, use the pH up.

Table 9.5: pH for some vegetable crops in aeroponics

Crops	pH
Cucumber	5.8-6.0
Lettuce	5.5-6.5
Onions	6.0-7.0
Potato	5.0-6.0
Spinach	5.5-6.6
Tomato	5.5-6.5
Carrots	5.8-6.4

Light and temperature

Sun light replacement is very important. It can be replaced by the appropriate fluorescent tubes Intensity.15000-20000 lux – for vegetative growth.35000-40000 lux – for flowering and fruiting. The optimum temperature for all plants is between 15 ° C and 25 ° C. But this state can be given by ventilation and exhaust air conditioning. The required approach can be selected according to the plant's necessity. The temperature within the growing chamber (which needs to be insulated) can be precisely regulated by heating or cooling the nutrient until it is misted into the root region. Depending to how harsh the aerial atmosphere is, crops with nutrient solution heating or cooling can only be grown in aeroponics, which is more difficult to do in other hydroponic systems where the solution easily cools / heats as it passes through the system. In Singapore, cooling the aeroponic solution by 10-150C below the average temperature of the air ensures that cooler seasonal crops can be grown at a temperature shift in the aerial climate.

Misting frequency and nutrient reservoir

Aeroponic systems can mist the root system continuously or intermittently, and both methods work well, as water logging and oxygen starvation are not a problem in aeroponics. The biggest benefit of temporary aeroponics systems is the decrease in operating costs, as the pump is only on for a brief amount of time, but the roots are also stored in a food, moisture and oxygen-rich atmosphere during mistings. Since aeroponic systems need larger pumps with higher energy requirements than other hydroponic systems, this saving is an significant benefit. Often look for systems that either connect light to misting frequency or have the potential to programme in a large number of misting cycles every 24 hours. As a general rule, the netting duration of 1-2 minutes of netting followed by 5 minutes of netting guarantees that the root structure does not dry out in certain conditions.

Aeroponics systems are also classified into those which provide a separate nutrient reservoir and pump the nutrient up into the root chambers and those which are all stored in a chamber and a nutrient reservoir. The simpler aeroponic systems are spraying the nutrient up from the tank in the bottom of the root chamber, where it drips back down after the root system is misted. By the time the plants mature, the root system has always evolved into a nutrient solution deposited at the base of the chamber, and blockages may occur. Larger aeroponic devices return the nutrient to a different nutrient pool during misting.

Aeroponics working method

in the aeroponic system, young plants can either be grown as seedlings using specially built lattice pots or cutters can be inserted directly into the aeroponic system for rapid root forming. Lattice pots allow the root system to grow into an aeroponic chamber or canal where it is periodically misted with nutrients. There is a high success rate with plant cuttings rooted in aeroponics-in fact, this approach has been commonly used as a root growth testing technique for many difficult-to-propagate plant species. The cutting base is provided with high levels of oxygen and moisture in a damp atmosphere that avoids desiccation and accelerates root development.

If a young plant has been developed in the aeroponics system, the root system grows rapidly in the chamber or path. What is critical at this point is that the ideal size of the droplets is preserved inside the device for optimum efficiency. There is a wide variety of aeroponic nozzles, so it is very straightforward to pick a droplet size variety that better matches the plant and the device used. Spray droplets of less than 30 microns appear to stay in the air as a 'fog' and are not quickly consumed by the roots. The average droplet size range for most species of plants is 20-100 microns. Within this size, the smaller droplets saturate the air, retaining the humidity levels within the growth chamber, the larger droplets of 30-100 microns make the most contact with the roots, whereas any droplets of more than 100 microns appear to fall out of the air before producing any roots.

The concepts of aeroponics are focused on the possibility of growing vegetables whose roots are not introduced into a substratum or soil, which are placed in containers filled with streaming plant nutrients. In these barrels, the root can find the best position for the best oxygenation and moisture content. These conditions allow for improved assimilation of plant nutrients in a more controlled manner, resulting in faster growth of the cultivated plant. Plant nutrition is given in a closed circuit. As a result, the intake is reduced to only the amounts absorbed by the plants, allowing for savings in power. For example: growing a kilogramme of tomatoes using conventional land takes 200 to 400 litres of water, hydroponics requires about 70 litres, aeroponics needs just about 20 litres.

Crop Production

Like hydroponic systems, aeroponics is also likely to be profitable producing high value crops.

Potato

It's C.B. Oh, Christie and M.A. Nichols., 2004 used aeroponics to mass-produce high quality seed potatoes and also to produce gourmet early (new) potatoes. When studying the early growth and evolution of potatoes in hydroponics, it soon became apparent that the root system had grown well ahead of the stolons to be formed by the tubers. The key to tuber development in aeroponics is to monitor the initiation of tubers. This can be done by using an intermittent irrigation system or by briefly stressing the vine. One of the benefits of the method is that coordinated tuber output enables a very large number of relatively standardised tubers to be produced at a single time.

The techniques to quickly multiply seed tubers, such as tissue culture in combination with hydroponic and aeroponic systems, have been attempted by Irman et al. in 2012 to alleviate the problem of lack of high quality seeds. Chang *et al.*, 2012 found that interrupting the supply of nutrients at the stolon growth stage dramatically enhances root operation, restricts stolon growth, and eventually induces tuber initiation. Non-tuberizing conditions, such as high temperatures and late-season cultivars, thus favour the use of this nutrient interruption technique, especially for the production of seed potatoes in hydroponics. Aeroponics may be an effective method for the development of potato minitubers. Up to 800 minitubers / m^2 could be generated in our system, using only 60 plants / mt square with repeated harvests per week. Minitubers thus developed were healthy, free of pathogens, and their physiological activity in field conditions was close to that of hydroponics. The International Potato Center (CIP) has recently produced and encouraged the development of mini-tubers based on a novel, rustic and publicly accessible aeroponics system. Results have shown that the aeroponics system is a feasible technical option for the potato mini-tuber production portion within the potato tuber seed system, which produces more tubers with a high tuber yield. Plants once washed by meristem cultivation and tuber induction under the aeroponics method, yield high-quality potato seed tubers easily free from pathogen infection, and the multiplication of potato tubers was simple and immediately planted in the main area. As a result, the aeroponic method has the ability to increase profits and reduce the cost of growing quality crop, rendering it more available to growers in developing countries where potato production is highly limited by the use of low quality seed tubers.

Yams

Aeroponics technology should be considered as an important way of spreading of yams. The two D genotypes. The rotundate and the D. Alata was successfully propagated using both pre-rooted and fresh vine cuttings. Results of these experiments have shown that vines from five-month - old plants were successfully rooted (95%) in aeroponics within 14 days. An average of 83 per cent (range of 68 per cent-98 per cent) of the rooting of the vine cutting was recorded for the five genotypes used. Genotypes behaved differently in aeroponics science for the development of mini-tubers. Yam mini-tubers harvested from aeroponics ranged from 0.2 g to 110 g based on the genotype, the age of harvest and the content of the nutrient solution. However, this mechanism is vulnerable to extreme heat, and caution must be taken to control temperatures.

Lettuce

Demar *et al.* (2004) studied the influence of light-dependent nitrate application on the growth and yield of aeroponic lettuce. He and Lee, 1998 observed that the growth of the shoot and root and photosynthetic reactions of the three cultivars of Lettuce (*Lactuca sativa* L.) confined to different root zone temperatures and growth irradiance under tropical aerial conditions was better than control for aeroponic crops. Luo *et al.*, 2009 have developed a very good method of generating hearty lettuce in the tropics using aeroponics and root refrigeration. It was also noticed that the effects of increased CO_2 root zone and air temperature on photosynthetic gas exchange, nitrate absorption and overall decreased nitrogen content in aeroponic lettuce plants were observed.

Leafy vegetables

The commodity production, total phenolic, total flavonoids and antioxidant properties were related in the different leafy vegetables / herbs (basil, chard, parsley and red kale) and fruit crops (bell pepper, cherry tomatoes, cucumber and squash) produced in aeroponic and field systems. Average increases of around 19 per cent, 8 per cent, 65 per cent, 21 per cent, 53 per cent, 35 per cent, 7 per cent and 50 per cent in yields were reported for basil, chard, red kale, parsley, bell pepper, cherry tomatoes, cucumber and squash, respectively, when grown in aeroponic systems compared to those grown in soil. Antioxidant properties of these crops were tested using 2, 2-diphenyl-1-picrylhydrazile (DDPH) and cellular antioxidant (CAA) assays. In general, the analysis indicates that plants grown in the aeroponic environment had higher yields and comparable phenolic, flavonoid and antioxidant properties relative to those grown in the soil. In aeroponics, the production of biomass was based on the density of the plants under investigation. The highest yield was achieved with a maximum density of 100 plants per m2. It has been shown that the length of the stems and the leaf area have both been impacted by the density of the plant. The vitamin C content was highest in all aeroponic herbs, while the essential oil content was highest in the Holy Basil and Perilla herbs grown in the substrate. The carotene content of water spinach was the highest in substrate community.

Advantages

» **Less fertilizer**-Since all the nutrients are stored, they do not end up in groundwater or fall too far into the soil to be of any use.

» **Less water**-Very important for space exploration for others in arid climates. Many of the water wasted in conventional gardening comes from the water that evaporates out of the soil. The rest of it all sinks past the roots, and the plants never get a chance to drink.

» **More cost effective**-Because less nutrient solution is required compared to hydroponics, the cost of running an aeroponic garden is smaller than that of running a hydroponic garden. There are also less moving parts and more complex structures involved.

» **Minimized disease damage**-Because plants are isolated from each other and do not share the same soil, illness in one plant has a much smaller risk of spreading to the rest of your plants.

» Faster and healthier growth as it has adequate oxygen (in the root region) The improved harvest rate is 45–70 per cent faster than traditional farming techniques.

» Experiments have shown that plants developed via the aeroponic system have increased flavonoids.

Disadvantages

» More costly for long-term production.

» Ordinary farmers will fail to handle all these advanced resources. Mister spray heads may also have a propensity to clog and not generate fog when appropriate.

» Most customers assume that aeroponic plants are not as nutritious as other cultivated plants.

» Maintenance of the aerospace farm is very expensive.

Conclusion

Aeroponics cultivation helps plants and crops to thrive without the application of a pesticide and thereby free from disease. Crops can develop in a safe natural fashion, since the aeroponic mechanism is somewhat close to the environmental conditions of nature. Aeroponics is carried out in the air combined with micro-droplets of water, almost any plant can expand to maturity in the air with an ample supply of carbon dioxide, water and nutrients. Aeroponics helps save water, soil and nutrients, so the aeroponics method is the way ahead, making the production of crops simpler. Aeroponics proved to be a highly viable process for generating both aerial and root ingredients as raw materials for the herbal dietary supplement and phytopharmaceutical industries.

Future Prospects

Aeroponics has the ability to increase productivity and reduce costs relative to traditional methods or other soilless hydroponics (water growth) approaches. Aeroponics successfully uses the vertical space of the greenhouse and air humidity equilibrium in order to maximise the growth of roots, tubers and foliage. Commercial processing of aeroponic potato seed is now underway in Korea, China and India. The technology has been used effectively in the Central Andean Region of South America since 2006. At the facility of the International Potato Center in Huancayo, Peru, yields of more than 100 minitubers / plants were obtained using relatively simple materials. Aeroponic technology is being studied in a number of African countries for the development of potato minitubers, and existing plans are underway to integrate aeroponic into the potato seed systems of certain sub-Saharan African countries.

Chapter - 10

Plant Growth Regulators for Hi-tech Horticulture

Some substances affect growth quite miraculously. This is referred to as hormones in early literature. Hormone is a Greek word derived from hormao (Upu w), which means to enhance or stimulate. Plant Bioregulators (PBRs), both natural and synthetic, have an impact on plant growth and development processes. Of such, phytohormones may be grouped into five classes, namely Auxins, Gibberelins, Cytokinins, Ethylene and Abscissic acid. The application of bioregulators involves seed germination, vegetative propagation, growth regulations, flowering, flower/ fruit thinning, fruit drop control, fruit development, improved fruit quality and pre-and post-harvest fruit crop management. Phytohormones are endogenous compounds which are naturally formed in plants, which regulate growth or other physiological functions at a location distant from its place of origin and are found in exceedingly limited amounts. Growth hormone as a "substance that is synthesized in specific cells and transmitted to other cells where the developmental cycle is influenced in extremely limited amounts." A hormone has been described as a substance that is generated in one part of a living organism and transmitted to another part to control any behavior. It was called a 'chemical messenger.' The term hormone, however, is quite popular and widely used. It is intended for an organic material that is synthesized in one tissue and migrates to another tissue of the plant where only limited amounts influence growth. In addition to these five groups of plant hormones, brassinosteroids, jasmonic acid, oligosacchrids, dormin, florigen and fusicoccin, etc., physiological activity has been reported. However, various types of substances influencing plant growth are now identified and can be narrowly categorized as growth-enhancing and growth-retarding substances or naturally occurring growth substances and synthetic growth substances. Growth-regulating compounds can also be categorized in the following groups.

i.   Auxins, e.g. Indole acetic acid (1AA)

ii.   Gibberellins, e.g. Gibberellic acid (GA)

iii. Cytokinins, e.g. Kinetin, Zeatin etc

iv.   Ethylene, e.g. Ethylene

v.   Dormins, e.g. Abscissic acid (ABA), Phaseic acid, Xanthoxin etc.

vi.   Flowering hormones, e.g. Florigen, Anthesin, Vernalin

vii. Miscellaneous natural substances, e.g. Cyclitols, Vitamins, Phytochrome, Traumatic substances etc.

viii. Phenolic substances, e.g. Coumarin.

ix.   Synthetic growth retardants, e.g. CCC, AMO 1618, Phosphon D, Morphactins, Malformins, Maleic hydrazide (MH) etc.

x.   Miscellaneous synthetic substances, e.g. Synthetic auxins, Synthetic cytokinins etc.

Auxins

Auxins are defined as an organic compound characterized by its low concentration capacity (below 10^{-3} M or 0.001 M) to induce elongation in the stem cells and inhibit the elongation of the root cells. These are similar to Indole acetic acid (IAA) in physiological action. They may and generally have an effect on other processes other than elongation, but elongation of the shoot cells is considered critical. The usage of synthetic auxins in horticulture can be linked directly to the normal function of IAA in the fruit. Compounds such as a-naphthalene acetic acid (NAA) are generally used because they resemble IAA in action but are resistant to degradation by plant enzymes.

Characteristics of Auxins: Auxins are characterized by the following characteristics-

»   Polar translocation (from peak to horizontal movement),

»   Apical bud dominance,

»   Variable behaviour of root and shoot growth,

» Open Root Initiation,

» accommodating the delay in abscission,

» Sympathize with the separation of the xylem elements.

Chemical nature of auxins

Chemically, it has been recognized as auxentriolic acid ($C_{18}H_{32}O_5$). In 1934, Kogl, Erxleben and Haagen-Smit extracted auxin b, auxenolonicacide ($C_{18}H_{30}O_4$) from corn germ oil. Later, another auxin was also isolated from human urine and referred to as heteroauxzin, which was recently called indole-3 acetic acid (IAA, $C_{10}H_9O_2N$). Auxin *a* and *b* are hardly known from the plant, but IAA is the main naturally occurring auxin to all higher plants and fungi. It also occurs in human urine, particularly in individuals with a deficiency in niacin or nicotinic acid. Indole-3-acetaldehyde, indole-3-acetonitrile, indole-3-ethanol, 4-chloro-IAA are also found in plants.

In addition to naturally occurring auxins, a variety of other substances with similar properties have been identified. Some of them are indole butyric acid (IBA), α and α naphthalene acetic acid (NAA), 2,4-dichlorophenoxyacetic acid (2,4-D), 2, 4, 5-trichlorophenoxyacetic acid (2,4, 5-T), indol propionic acid (IPA), naphthoxyacetic acid, 4-chlorophenoxyacetic acid, 2,4, 6-trichlorobenzoic acid, 4-amino-3,5,6-trichloropicolinic acid, etc.

Site of auxin synthesis and translocation: For example, the known active sites of auxin synthesis include the tip region (especially young apical bud growing leaves), the tip of the grass or cereal seedling coleoptiles and the development of embryos, the development of fruits, the enlargement of tissues, etc. Auxiliary synthesis occurs quickly in the presence of light in the green leaves than in the dark. Auxin is transported basipetally, i.e. it moves from apical to basal. The acceleration is very quick, around 1 to 1.5 cm / h (0.1 to 0.2 cm / h in the root). Recent experiments have shown that the auxin transport mechanism in plants is an overall successful transport device. Many of the arguments in support of it can be listed as follows:

The velocity of auxin transport (1 to 1.5 cm / h) in stems and coleoptiles is approximately ten times faster than diffusion. Metabolic inhibitors are less effective in transmitting auxins relative to poisons. 2,3,5-triiodobenzoic acid and naphthylthamic acid are very effective. Transportation of Auxin relies on aerobic metabolism.

Auxins can shift toward a concentration gradient. It has been noted that there is a good correlation between the structural requirements for auxin activity and for auxin transport. This is also accompanied by the saturation impact of the transport site between 0.1 and 1 mm per volume of tissue. Auxins are used for a wide variety of horticultural purposes, including:

» Promote the rooting of cuttings (e.g. Rootone). Until planting, the base of the cutting is dipped in a powder comprising NAA or indolebutyric acid (IBA).

» Induction of flowering in pineapple (actually triggered by the auxin-induced development of ethylene). NAA is commonly regarded as an auxin.

» Improved fruit collection and pericarp development in the absence of fertilization. Preventing the decline in pre-harvest berries.

» Apply herbicides of the Auxin (e.g., 2-4-D).

» Preventing germination through repression of buds and extended dormancy or regulation of flowering.

» Preventing the dropping or abscission of leaves and thinning of compressed fruits.

Gibberellins

Since the discovery of GAs in the 1920s, a substantial body of information has arisen as to the function of these hormones in plant growth and development. As these compounds and analogs became usable, either by fermentation or by chemical synthesis, a number of potential practical applications became established. Gibberellins have been used primarily to manipulate production practices and to ensure the quality of high-quality specialty crops such as grapes, citrus, cherries and apples. Increasing consumer demands for a greater variety of quality fruit that create new opportunities for expanded use of GAs (i.e. tropical fruits, thinning and sizing of wine and new table grape varieties, and as components of integrated pest control programs for post-harvest fruit quality). Notwithstanding extensive work, GAs have not been shown to be effective in large agronomic crops. Nonetheless, an improved awareness of the interrelationship between GAs and other PGRs, in combination with the production of new GAs or GA-mimics, can deliver the specific activities required to enter agronomic markets.

However, following comprehensive work with GAs, the amount of industrial uses has fallen short of their anticipated effect and development in agriculture. New prospects can also emerge from new research and development with GAs to satisfy potential demands for better food quality and growth.

Characteristics of gibberellins: GA is distinguished by the following features.

> » Prevention of genetic and physiological dwarfism,
> » Breaking of dormancy,
> » Induction of flowering in long day plants,
> » Increase of amylase activity,
> » Substitute for chilling effect.

Chemical nature of gibberellins

Gibberellins are cyclic diterpenes. Chemically these are represented by molecular formulas such as $C_{19}H_{24}O_6$ (GA_1), $C_{19}H_{26}O_6$ (GAz), $C_{19}H_{22}O_6$ $(GA_3$ or Gibberellic acid), $C_{19}H_{24}O_5$ (GA_4), etc. They contain ɤ-lactone ring on a cyclohexane ring. The existence of hydroxyl groups and carboxyl groups is produced by an esterified product. The OH community is growing their biochemical behaviour. Although the COOH party has reduced response.

Approximately 120 specific gibberellins have been reported in higher plants and the Gibberella fungus, but only two (GA_3 and a mixture of GA_{4+7}) are accessible by fermentation of the fungus *Fusarium monilifom (Gibberella fujikuroi)* in a sufficient quantity for detailed testing. The nature of GA chemistry precluded fully synthetic distribution paths. With the complexity of physiological processes in which GAs are active, it is not shocking that their functional implementation covers the entire spectrum of plant growth and production. However, several of these applications are, in fact, small outlets for high cash value horticultural crops. Commercially, gibberellins are derived from fungal cultures and are refined natural products that are added to plants.

Site of synthesis and translocation: GAs are synthesized from mevalonic acid in young leaf tissues (precise location uncertainty), root tip, and developing seeds (embryo). Gibberellins travel quickly in both directions and in all tissues, including phloem and xylem. GA travels in the same direction as the carbohydrate translocation mechanism moving at a comparable pace (5 cm/ h). As compared to auxin, it is inactive, non-polar and diffuse.

The following are the main impacts on plants:

» **Stem growth:** GA_1 induces hyper elongation of the stem by causing both cell division and cell elongation. This produces tall, rather than dwarf, plants.

» **Bolting in long day plants:** GAs cause elongation of the stem in reaction to long days.

» **Induction of seed germination:** GAs can cause germination of seeds in some seeds that usually need cold (stratification) or light to induce germination.

» **Enzyme production during germination:** GA promotes the development of a variety of enzymes, especially a-amylase, in the germination of cereal grains.

Gibberellins are used for a wide variety of horticultural purposes, including:

1. Enhanced production of seedless grapes. Bigger, more even bunches of larger fruit are made. Among other effects, gibberellins cause the lengthening of the peduncle (stalk) by attaching each grape to the cluster, thus allowing the formation of larger grapes. Virtually all the grapes on the market are already labeled with gibberellins.

2. Treatment of oranges to avoid senescence of the rind, to enable longer preservation on the fruit, and thereby to prolong the marketing time.

3. Enhanced flower bud development and increased fruit production in cherries.

4. Improvement of fruit setting in apples and pears, particularly in weather conditions that are bad for setting.

5. Replacement of the cooling criterion in situations such as: (1) crop induction for seed development (radish), (2) improved elongation (celery, rhubarb) and (3) early flower production (artichokes).

6. Plant of hybrid seed of cucumber. Most of the high-producing cucumbers are F_1 hybrids. GA sprays stimulate the growth of male flowers on cucumber plants, which usually generate only female flowers. The seed comes from all female plants with a particular strain is then specifically hybridised.

7. Improved output of barley. The application of GA to germinating barley during the processing of beer improves the development of a-amylase such that more malt is extracted more rapidly. Because malt is the raw material for fermentation, this method allows it easier to create more beer.

Cytokinins

The word cytokinin is commonly used as a general name for a drug that facilitates cell division and performs certain development regulating roles in the same manner as kinetin. This compound was known as 6-furfuryl aminopurine and was referred to as kinetin. The most popular cytokinin base in plants is zeatine, dihydrozeatine and triacanthene. Several substances have been found that show cytokinins like activity. Common forms of this form include 6-aminopurine (adenine), benzimadazole, 6-benzyladenin, 1-benzyladenin, etc.

Characteristics of cytokinins: Cytokinins are characterized by the following characteristics:

» Initiation of the division of cello,

» Senescence delay (Richmond-Lang effect)

» Correct use in tissue culture

» Counteract apical bud supremacy,

» Induces flowering of plants on a limited day.

Chemical Nature of Cytokinins

Cytokinin was found to contain carbon, hydrogen, nitrogen, and oxygen in the $C10H9N5O$ ratio. Cytokinin consists of an extra-carbon furan nucleus present either as a side methyl on the furan ring or as a methylene radical between the furan and the purine nucleus.

Site of synthesis and translocation: The root tip is an essential synthesis site. However, the growth of seeds and evolving tissues is also the source of cytokinin biosynthesis. Cytokinins have so far been extracted from coconut milk, tomato juice, flowers and fruits of Pyrus malus, fruits of Pyrus communis (pear), Prunus cerasifera (plum) and Lycopersicum esculentum, changing tissues of Pinus radiate, Eucalyptus regnans and Nicotiana tabacum, immature fruits of Zea mays, Juglans sp. And this is Musa sp. Male gametophytes of Ginkgo biloba, fruitlets, embryos and endosperms of Prunus persica, seedlings of Pisum sativum, root exudates of Helianthus annuus and tobacco tumor tissues. Cytokinins travel upwards (root to shoot), possibly in the xylem current, there are also records of their basipetal movement in the petiole and isolated base.

The function of cytokinins is therefore widely known and there is little data to establish any molecular point of function. The main consequences are:

1. **Cell division:** Effects of cytokinins cause cell division in the presence of auxin in tissue culture. The appearance of cytokinins in tissues of rapidly dividing cells (e.g. fruits, shoot tips) suggests that cytokinins will actually serve this role in the plant.

2. **Morphogenesis:** Cytokinins facilitate the activation of shoots in tissue culture.

3. **Growth of lateral buds:** Cytokinins can trigger lateral buds to be released from apical supremacy.

4. **Leaf expansion:** Resulting mainly from cell enlargement. This is probably the mechanism by which the total leaf area is adjusted to compensate for the extent of root growth, as the amount of cytokines reaching the stem will reflect the extent of the root system.

5. Cytokinins delay the senescence of the leaf.

6. Cytoldnins can enhance the opening of the stomach in some animals.

7. **Chloroplast development:** The use of cytokinins results in the accumulation of chlorophyll and promotes the conversion of leukoplasts into chloroplasts.

Big consumer applications for cytokinins benefit from their potential to inhibit senescence and preserve greenness. Artificial, chemically active cytokinin, benzyladenin, is the primary agent used. Holly preparation for holiday decorations requires it to be extracted several weeks before use. Post-harvest sprays or dips are now available to prolong the shelf life of green vegetables such as asparagus, broccoli and celery.

Abscisic acid

The term abscissic acid (ABA) is very undesirable. The first name given was "abscisin II" because it was thought to control the abscission of cotton bolls. Almost at the same time, another group called it "dormin" for the purported role in bud dormancy. By a compromise, the name abscissic acid was coined. It now appears to have little role in either abscission or bud dormancy, but we are stuck with that name. Though exogenous applications that inhibit plant production, ABA tends to serve as a promoter (e.g., protein synthesis in seeds) as an inhibitor, and a more transparent approach towards its overall function in plant development is required. It also acts as a stress hormone that helps plants cope with adverse environmental conditions.

Characteristics of abscissic acid: Abscissic acid is characterized by the following characteristics:

» Abscissic acid associates with IAA, cytokinin and GA in a number of physiological behaviours.

» Regulation of the orientation of the guard cells and of the stomach operation.

» Beginning of senescence in plants

» Decreased rate of photosynthesis and increase of nucleic acid.

» Abscissic acid causes flowering during long days.

Chemical nature of abscissic acid

Abscissic acid, chemically 3-methyl-5-(1'-hydroxy-4-oxo-2' 6' 6', = trimethyl-2'-cyclohexen1'-yl) cis, trans-2,4-pentadienoic acid is one of the most common and significant metabolites in plants. Abscissic acid has an asymmetric carbon atom and may also act as either (+) or (-) enantiomers. Abscissic acid naturally occurring is usually in (+) form (2 *cis*-ABA) and is more active than (-) form (2-trans-ABA). Recent research has shown that 2-trans-ABA is not itself biologically active but is converted into *cis* in the presence of ultraviolet radiation. A variety of synthetic plant growth retardants have been produced which induce plant dwarfing without any harmful effects. The main compounds known to be effective include nicotinium, phosphonium, choline, hydrazine, pyrimidine and triazole compounds.

Site of synthesis and translocation

Abscissic acid is synthesized from mevalic acid in roots and mature leaves, particularly in response to water stress. Seeds are often high in abscissic acid, which can be introduced from or synthesized from the seeds. Abscissic acid is extracted from roots in the xylene and from leaves in the phloem. There is some indication that abscissic acid can migrate to the roots of the phloem and then transfer to the xylene shoots. Both plant components are believed to be synthesized. The important effects of abscissic acid are as follows:

1. Abscissic acid induces bud dormancy in plants.

2. Abscissic acid accelerated abscission of leaves in different horticultural plants, particularly under stress conditions.

3. Abscissic acid induces flowering over long days.

4. Abscissic acid triggers ethylene production.

5. The plant senescence is stimulated by abscissic acid.

The rise in the production of abscissic acid is an sign of senescence, marked by depletion of chlorophyll, reduced photosynthesis and improvements in nucleic acid and protein metabolism. In this way, abscissic acid may be viewed as a chemical that signals the beginning of senescence in plants.

Ethylene

Ethylene (C_2H_4) is a gas at temperatures at which the plant will work. Ethylene is present in plants and functions as a chemical. It is well established that ethylene is provided by flowers, leaves, stems, roots, tubers and seeds. The highest rate of development of ethylene happens during the normal breathing cycle just before senescence in many fleshy plants, accompanied by a rise in the breathing rate called a climacteric rise. Here it might be assumed that the production of ethylene may be closely connected with the ripening cycle of the fruit. Gas ethylene is synthesized from methionine in several tissues to react to stress. It does not appear to be necessary for normal vegetative development. It is the only hydrocarbon that has a marked effect on plants. Some tissues synthesize ethylene in response to stress. In specific, it is synthesized in senescent or ripening tissues.

Characteristics of ethylene: Ethylene is characterized by the following characteristics:

»   Ethylene is just the gaseous source of the plant hormone

»   Ethylene induces rapid abscission of leaves and seeds.

»   Ethylene is a ripening agent

»   Eethylene raises the number of female flowers.

Chemical nature of ethylene

Chemically, ethylene is the simplest of all plant-growing substances. Other olefin molecules, such as acetylene and propylene, may produce ethylene as an action, but ethylene is 60-100 times more active than propylene. They are small molecular weight hydrocarbons with an unsaturated bond next to the terminal carbon atom. In recent years, the development of the ethylene release agent-2-chloroethyl-phosphoric acid sold as Amchem 66-329, CEPA and etherel has generated substantial horticultural concern. Ethephon exerts its influence by acting as a decomposition product in plant tissue and by releasing ethylene near the site of action. The new product is successful at concentrations of 100-500 ppm in aqueous solution.

Site of synthesis and translocation

It is well established that ethylene is released by flowers, leaves , branches, roots, tubers and seeds. The highest rate of production of ethylene exists in many carbohydrate fruits during the prolonged respiratory time just before senescence. As a gas, ethylene travels through diffusion from its synthesis source. Ethylene is a tiny molecule that makes ready passage across plant tissues by diffusion. The passage through tissues may be through openings in the air and thus connected to tissue porosity. Its solubility in water and even its lipophilic nature would make it possible to slip across the cell membrane, or even activate the operation of the permease mechanism.

The important effects of ethylene are:

1. Ethylene has a wide range of uses, but its gaseous nature prevents it from being used in unenclosed areas. It also influences flowering (in pine apples), apical bud superiority, stem and root growth reduction, petal discoloration, abscission, maturity, freezing and disease tolerance. Ethylene itself may be used to improve the ripening of fruit such as bananas in storage after being delivered in unripe form, which is of great advantage, because green bananas are durable and do not bruise or spoil easily. The tender of ripe bananas can then be transported safely from the nearby warehouse to the market. Recently, an ethylene-producing liquid product, 2-chloroethyl-phosphonic acid (widely known as Ethrel or Ethephon) has been commercially launched. This compound is sprayed onto the plant at a slightly acidic pH. It breaks down to release gaseous ethylene when it enters the cells and encounters a cytoplasm at about neutral pH. Numerous commercial applications have been produced for this material, mostly in relation to the natural effects of ethylene:

2. The most important commercial application involves increasing latex production in rubber trees in South East Asia. When the rubber tree is "tapped," the latex runs for a certain period of time before the seals are

removed and the discharge ceases. Ethephon slows the curing of the wound such that the fluid release lasts for a prolonged period of time, resulting in more injection and fewer rubbing.

3. Ethylene accelerates the abscission of seeds, bulbs and fruit.

4. Enhancement of uniform ripening and coloring of fruit. It has been seen to be of specific importance in the field of tomatoes harvested by system at a single time.

5. Acceleration of the abscission of fruit for mechanical harvesting. This provides a potential niche for use in a large variety of fruits such as strawberries, cherries and citrus.

6. Promotion of the development of female flowers in cucurbits (cucumber, squash, melon) in order to increase the amount of fruits produced per vine.

7. Promotion of flower initiation and regulated maturation in pineapples.

Use of Plant Growth Regulators in Hi-Tech Horticulture

Plant growth is used in hi-tech horticulture to improve yields, improve quality and promote processing, together with genetic engineering, are complementary technologies in horticulture to maximize quality production and optimize the use of limited resources. While plant growth regulators have been used in horticulture for as long as crop protection chemicals have been used, their effect to date has been relatively small and their use is limited to specific crops.

However, most popular applications for plant growth regulators in high-value horticultural crops are not for compounds that directly increase crop yield, either by increasing total organic yield or by increasing the harvest index. Rather, compounds that provide economic benefits by enhancing crop quality or helping to manage crops more effectively are more common. For eg, gibberelic acid is used to reduce the occurrence of physiological rind disorders in citrus and daminozide (N, N, dimethylaminosuccinamic acid) in apples to promote fruit color growth. Both treatments improve the output of the crop but do not automatically increase the yield.

Plant growth regulators that help hi-tech horticulture crop control fell into many categories:

1. Mechanical harvesting of crops will greatly reduce the cost of output. Compounds that minimize the fruit removal force in certain situations,

such as sweet cherry processing, have enabled the invention and usage of mechanical harvesting machinery, which was not especially successful on its own because of the force needed to extract weakened trees from the field.

2. Plant growth regulators are used to control growing dates in a wide variety of crops. Ethephon is used to stimulate and optimize the ripening of tomatoes prior to a single automated harvest. Gibberellic acid is used to extend the lemon growing season by avoiding senescence of the rind.

3. Plant growth regulators can be used as direct substitutes for hand labour in crop processing activities other than harvesting. Naphthalene acetic acid (NAA) is used in apples to reduce the excess fruit collection which will otherwise result in other tiny fruits.

4. Propagation of the plant by seed germination, cutting, layering, grafting and tissue culture.

5. Preventing fruit declines and growing fruit collection.

6. Induction of artificial seedlessness.

7. Regulation of post-harvest growth (breeding and root forming at the time of storage).

8. Inhibition of development of nursery stock buds in storage or momentarily halting production after planting

9. Regulation of flowering and fruit scale as in pineapple, pomegranate, orange, etc.

10. Thinning of flowers and fruits include mangoes, apples and peaches.

Chapter - 11

Plasticulture for
Hi-tech Horticulture

Plastics are used in agriculture, horticulture, water management, food grain storage and related areas. This is known as the use of plastics in the agriculture of plants and animals.

Plastic products themselves are sometimes and commonly referred to as *ag* plastics. Plastics include soil fumigation material, irrigation drip tape / tubing, nursery pots and silage bags, but the term is more widely used to define all sorts of plastic plant / soil coverings. These covers range from plastic mulch film, row covers, high and low tunnels, to plastic greenhouses.

A wide range of plastic materials and end products are used in plasticulture applications for water conservation, irrigation efficiency, crop and environmental protection, as well as final product storage and transport. Plastics can play a key role in the conservation of energy. They require minimum energy for production and conversion to finished products. Plastics have clear advantages over conventional materials such as higher strength / weight ratio, superior electrical properties, superior thermal insulation properties, excellent corrosion resistance, superior flexibility, water, gas, etc. impermeability, chemical resistance and less friction due to smoother surfaces. As a result, plasticulture applications are one of the most useful indirect agricultural inputs that promise to transform Indian agriculture and bring about the Second Green Revolution. It is used in water management, drip irrigation, controlled environmental agriculture and -LEPD film mulching.

By cutting costs and saving time, plastics make agriculture more productive and efficient. They also conserve valuable natural resources, such as water , nutrients, fossil fuels and many forms of energy, especially sunlight. In some cases, plastic can be used to reduce sunlight and retain moisture, especially in arid regions. It also reduces competition from invasive weeds and insects.

Plastic culture farming has evolved dramatically in recent years through experimentation, field trials and working relationships between commercial growers, nurseries and experimental stations. The term plasticulture refers to planting and growing techniques which may or may not involve the use of any plastic material. The old idea that plastic culture must mean plastic mulch has changed. For example , in the case of a progressive strawberry grower, plasticulture farming begins with raised beds of dual-row strawberries planted to stand with drip tape irrigation. Plastic mulch is a possibility. For all growers, there is no single best method.

Key benefits of plasticulture include: superior weed control, reduced soil moisture loss for nutrients, longer potential harvesting season, efficient and convenient harvesting, increased fruit size and yield and more marketable fruit.

Typical products used in plasticulture shall include:

1. Low tunnel, high tunnel, nursery house and greenhouse covers
2. Drip irrigation tape and tubing
3. Row and floating row covers
4. Bunk silo covers
5. Mulch films
6. Hay and silage bags and wraps
7. Baling twine and ties
8. Baskets and clamshells
9. Pails and barrels
10. Sun shades and screens
11. Fumigation films
12. Edible films
13. Plastic fittings and parts associated with many of these systems.
14. Nursery pots, trays, tray inserts, and flats
15. Degradable and biodegradable plastics

The National Committee on the Use of Plastics in Agriculture (NCPA) was founded in 1981 to encourage the use of plastics in agriculture. Its mission is to provide state governments with funding for research and training through its Plasticulture Development Centers (PDCs), oversee the introduction of plasticulture schemes and assist in technology transfer. The NCPA currently has 11 PDCs, most of which are located in State Agricultural Universities. Six more PDCs are being set up to serve the big countries.

Plasticulture Use in Hi-Tech Horticulture

Plastic mulch

Plastic mulches have been commercially used on vegetables since the early 1960s. Many of the early university work prior to 1960 was performed on the effect of color (black of transparent plastic film) on soil and air quality, moisture preservation and crop yields (Emmert, 1957). Most plastic mulches used in the United States are made of linear low or high density polyethylene. They are 0.012 to 0.031 mm thick, 122 to 152 cm across and 607 to 1463 m long on plates, depending on the thickness of the mulch. Linear high-density polyethylene is used to minimize weight and expense and is thicker than the same low-density polyethylene material. The plastic mulch is either slick or embossed with a diamond-shaped pattern. This method serves to limit the expansion and contraction, resulting in the loosening of the mulch from the raised bed. The raised bed is generally 10-15 cm high and 75 cm wide and has a slope of 3 cm from the center to the edge. The soil under the raised bed would warm up more easily in the spring, and would also release excess water from the center of the field, thereby holding the crops dry and stopping the quality of the crops from degrading. the product.

Many of the other advantages of plastic mulch include early growth, reduced weed pressure in the field bed, reduced soil evaporation, reduced soil splashing on crops, etc. Plastic mulch is now a part of the development of plastic culture. Plastic crop production often uses a number of different tools to increase production. The most commonly used products are plastic mulch and drip irrigation. Nevertheless, other items such as floating line covers, low tunnels, large tunnels, etc. are also included.

Fig. 11.1: Plastic mulch

Melons are one of the crops that also have a positive reaction to the use of plastic mulch. Black plastic mulch is more widely used, although yellow, white and other colors are available. Plastic mulch tends to raise surface temperatures early in the growing season, helping moist summer crops such as melons. Different mulches warm the soil in a number of ways. Black plastic heats the soil by close contact. The sun heats up the black plastic, and it touches the soil where the heat is transferred. Areas where there is a much slower gap between soil and plastic heat. The air gap is acting as an insulation. Before the heat is transferred to the soil, the air must be heated.

Clear plastic mulch warms the soil the best. Clear plastic enables the sun's rays to glow by the intense heating of the earth. Plastic serves to trap the heat in place. Soils under transparent plastic will be 5 degrees or more, cooler than bare soil in the same container. The downside of transparent plastic is that the sunshine that passes through the mulch often helps weeds to spread. A crop that easily covers the plastic will reduce the development of weeds. Or, some herbicides may be used underneath plastic to reduce weed growth.

Selective mulches wavelength

Another alternative is to use specific mulches for wavelength. Selective spectrum mulches aim to blend the strengths of both black and white plastic mulch. The theory is that the wavelengths that warm the soil move through the mulch, but the wavelengths that cause the plant to expand are absorbed by the mulch. These products fall in-between clear and black plastic in their soil heating properties. These will prohibit much of the weeds from developing, although there are a few plant types that can thrive under them. Such goods are also green or brown in colour.

Other mulch colours

Many experiments have looked at various shades, red, yellow, green, blue, etc., with specific crops. Few experiments have shown that various crops are better matched to other colors. The findings were complex. Some of the issues is that standard colors have not been developed. Red isn't just red! It's red, pink, rose red, and of course ... It's scarlet! White mulches are often used to help cool soils during hot times. To avoid weed growth under white mulches growers should try white on black mulch. There are also transparent mulches that can be used to better manage insects and diseases. Such materials are designed to confuse virus vectors like aphids. Such products can be successful, but only as long as they are not hidden by the crop canopy.

Commercial mulch

Many industrial mulch products provide photodegradable mulch broken down by sunlight. This forms of mulch had inconsistent outcomes owing to an unpredictable breakdown. They split at the moment they like. Other times, they could break down too early or too late. Also, the buried edges will not be broken down until they are exposed to sunlight. Research is also being undertaken to create a paper mulch. This would have been a big benefit, because it could be worked on the ground instead of having to be picked up after the growing season. The paper mulch has not been mastered so far. The product we tested was difficult to lay down, often tearing. Additionally, the coil is somewhat wider in diameter and thicker than the silicone sheet. Hopefully, a paper mulch company would soon be on the market.

As for all things, there are a few downsides to a plastic mulch. The greatest challenge is separation and disposal from the ground. Plastic can be difficult to pick up, and it's pretty dirty, which limits the ability to recycle. In addition to potential pesticide contamination. The landfill is the most popular method of disposal.

Although plastic mulch can inhibit weed development, it can also produce some challenging weed control problems. Weeds continue to spread into plastic cracks. Yellow nutsedge develops right through a green mulch. Weeds often appear to develop right around the edge of plastic where they hide from mechanical cultivation. Many herbicides can be used under plastic mulch, although some cannot. Growers also need to be careful about applying herbicides to row middles and putting herbicides on the mulch. Roundup stays on the mulch until it is washed away. When the young melon vines hit it, they'll pick up the roundup first. And, if it is added to the plastic mulch stream, it can wash it in the plant hole and ruin the melon transplant.

Weed control

One benefit of the plasticulture method is the outstanding weed management of dark plastic mulch and fumigants (where used). However, weeds can still be found in the hole of each plant, in the middle rows and along the edges of the field. Weeds sometimes flourish in barren soil at the ends of the rows or the edges of the field. Farmers allow weeds to germinate and destroy before planting. It is recommended that plastics be laid 30-40 days before planting. However, shorter cycles may also be a win. Herbicides are frequently used in line, but caution must be taken to insure that these products do not spill out of material and accumulate in cracks.

Weeds decrease the yield of melon, reduce the number of melons and reduce melon size. Weeds can delay maturity and can minimize efficiency. Severe plant infestations conflict with farming. Such adverse consequences of plants can be minimized with an effective weed control system.

The effective weed management program for melons shall include:

1. Identification of the major weed species in the area.

2. Usage of cultural activities to reduce weed growth and seed development.

3. Use of weed control practices to reduce weeds and kill weeds.

A sensible period to classify plant species in the field is the year before melons are grown. Notice, occasionally throughout the season, what weeds are present and whether they are dispersed throughout the field or clustered in a few areas. Notes on plant organisms can aid in the application of herbicides and other plant management methods for melon crops.

Cultural activities such as fertilization, irrigation and seed spacing have an effect on weed production. If these methods can be used to mitigate weed production, weeds can be easier to destroy. Weeds that control the escape will compete less with the crop and produce less seed. For example , by adding fertilizer and irrigation to beds only, rather than distributing it over the whole field, weed growth between beds may be can. Crop spacing is critical because the closer the rows and plants are together, the faster the canopy shadows the ground. A complete canopy severely reduces the amount of light small weeds received, limiting their growth and the production of seeds. Row spacing is also critical as it determines what tractors and cultivation equipment match between the rows.

Weed control practices include the use of herbicides, cultivation, plastic mulch and preparation of stale seed beds.

The stale seedbed system can provide some control of weed species that occur early in the season. In order to achieve adequate weed control, the seedbed method should be combined with other techniques. It is more successful when it comes to larger plantings, as there is room to encourage weeds to grow.

Opaque plastic or paper mulches successfully suppress weeds in and around the crop path.

Low and high tunnels

Low tunnels are suitable for the early growth of a variety of vegetable crops. These are built by mounting 6-foot long sheets of either cut or perforated plastic over wire hoops, which are placed about 4 feet apart. To avoid exposure to the wind, the cover is folded over the hoops (but not stretched) and the corners are covered with the dirt. Typically, the cover is placed over the plants for 3-4 weeks. Slits and perforations provide daytime ventilation, eliminating the need for manual ventilation. Low tunnels are commonly used in combination with black plastic mulch for plant control and better surface heating.

Frost protection from 2 to 4 °F is offered by the covers, but the main advantage is the boost given by the frequent rise in solar heating. Daytime temperatures may be excessive under cover if ambient temperatures are above 90 ° F. Under these circumstances, partial removal of the cover can be appropriate, particularly for tomatoes and peppers with open flowers.

Row covers, high tunnels and low tunnels will render earlier vegetable crops possible by generating a mini-greenhouse effect. The first row covers used were sturdy polyethylene sheets that needed protection and allowed daytime ventilation.

A number of products have been established to reduce the need for hand venting, including slit polyethylene covers including wire hoops; floating nonwoven sheets; white point-bonded polypropylene and polyethylene sheets with tiny pores. Row covers also help drive insect pests safe.

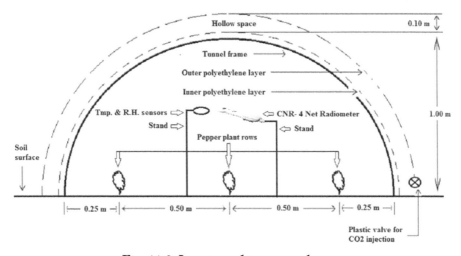

Fig. 11.2: Low tunnel type greenhouse
Source: Al-Kayssi and Mustafa (2016)

Fig. 11.3: High tunnel type greenhouse
Source: *https://blog.uvm.edu/*

Method of installation

1. Space hoops about 4 feet apart to prevent the coverings from sagging.

2. Apply covers for a peaceful day.

3. Attach the cover taut and protect the edges of the dirt.

4. Machines are available for installation on hoop and cover.

5. Installing the cover right around the edges of the black plastic mulch to avoid plant development under the sheet.

Advantages of slitted/Perforated row covers

1. Outdoor crops can be planted 10 days earlier than usual.

2. Best heat preservation with plain perforated covers and slit covers

3. Earlier maturity and higher yield

4. No manual ventilation needed

5. Good for tomatoes, onions, cucumbers, melons and several other vegetables.

Crops that are Suited for Slitted/Perforated row covers

1. Vegetables-hot season and cold season crops that are planted

2. Small fruit, strawberries in a matt group

3. Cut flowers

The results of the use of slit or perforated rows will be different for different crops, but in general all crops will benefit. The primary precaution is to provide extra ventilation when there are daytime temperature spikes. Drip irrigation and mulch is advised as part of the rising program.

High tunnel system may also be constructed for the protection of crops using plastics. High tunnels are another choice for growing vegetables in a plastic culture network. These may be used to prolong the rising seasons of spring and fall. Deep tunnels are filled by a thin sheet of polyethylene film. The use of high tunnels is common in many parts of the world, especially in Asia and the countries of Spain and Italy.

The Haygrove Tunnel System

Haygrove is a multi-bay framework introduced from Southern Europe. The idea was a positive one, but the tunnels required drastic changes to suit our needs. For the years, innovative systems were evaluated and strengthened on the farm before we had a low-cost, field-scale tunnel network in which tractors could be used and which would hold up to heavy winds. The effect on British soft fruit horticulture has been remarkable and has extended to other crops with over 4,000 acres of Haygrove tunnels in the last 5 years including small bananas, cherries, plums, roses, spices, ornamentals, onions, peppers and several high-value vegetable crops. Haygrove tunnels are already being used all over the world to help farmers generate safe stocks of good quality seeds, fruits and vegetables. North American growers are now seeing the success stories of strawberry, mango, plum, cut flowers and tomato farmers.

Fig. 11.4: The Haygrove Tunnel System

Table 11.1: Use of Haygrove tunnels and crop production

Crop	Benefits
Tomato	Increased yield
	Dramatic reduction in early and late blight infection (*Phytophthora* and *Alternaria*)
	Improved skin finish
Cherry	Increased yield
	Fruit allowed to size by safely delaying harvest.
	Losses due to cracking and rots virtually eliminated
	Better quality
	Larger, sweeter fruits

Raspberry	High quality late season production from fall bearing raspberries
Strawberries	Reliable supply to end customers and more efficient use of labour
Raspberries	Clear economic benefits: 14% yield increase, 34% increase in grade one yield, 2-3 weeks earlier start to picking, 31% more income

Runner control

plants are not permitted to be wrapped in fiber. Runners should plant only as the grower does. The usage of growth regulators can be seen as inhibiting the production of runners. Work and physiology of strawberry suggests the launch of the flower sprint. Research has found that monthly permits are better than one broad dosage.

Pest management

The plasticulture method must have a strong coordinated plan for pest management, disease control and plant control. In order to achieve effective insect control and disease control, it is necessary to use a sprayer that creates adequate pressure to enable pesticide sprays to penetrate and cover the whole field. This ensures that sprayers with pumps capable of producing more than 200 psi with correctly designed nozzles are required. It is necessary to use chemical sprays effectively and to monitor the targeted pest without harming the ecosystem. The usage of an Integrated Pest Management (IPM) method incorporating the use of disease-resistant crops with chemical and biological regulation, crop rotation and successful surveillance is suggested. Only herbicides approved for growing vegetables should be used within rows of mulched vegetable beds, as this should not be considered a fallow field. For gaps between garden beds, the usage of low-pressure sprayers, combined with shielded application of herbicides, is advised. This strategy would shield the mulched crop from herbicide, ensuring that no accumulation of herbicide in the planting hole can be harmed.

Reflective plastic mulches, such as silver mulch, have been found to conflict with the passage of aphides. Aphids are widespread carriers of virus diseases in different vegetable crops (e.g. watermelon mosaic virus II, which induces green streaks in summer squash and mottling and green streaks in yellow squash, melons and pumpkins).

Soil solarization and strip fumigation

The soil on which greenhouse or mulched crops are cultivated must be sanitized. Plastic mulches are used for chemical fumigants or as a barrier during soil solarisation.

In the row or strip application of a fumigant, the volume of material currently utilized per hectare may depend on the width of the row and be a proportion of the transmitted average. The soil temperature should be at least 10 °C and the soil should be well treated, clear from undecomposed plant debris and have enough moisture for seed germination. When the environment and soil are dry, the fumigant should be permitted to escape through the plastic mulch in 12-14 days. Fumigation is mainly used for nematode protection, but a multipurpose fumigant may also have effective control of soil-borne diseases. Solarization of soil is another way to manage plant pests. Solarization defines a hydrothermal process of soil purification that exists in saturated soil covered by mulch film (usually clear) and exposed to sunlight during hot summer months.

Windbreaks

The usage of windbreaks, whether permanent (trees) or annual (grain crops), is an significant part of the plastic culture development method, which is sometimes ignored. Windbreaks made up of strips of winter wheat, rye or barley are sometimes planted to shield young crop seedlings from prevailing winds. A mixture of permanent and annual windbreaks can change wind patterns and affect temperatures and other microclimate characteristics. Windbreaks can also act as shelter for both beneficial and harmful insects. To order to reach optimum performance, the grain strips will be planted to the fall. Strip of grain crop will be 3 to 3.5 m long (the width of the seed drill). Sufficient space should be left between the strips for five or six mulched beds of vegetables, each around 2 m long. Topdressing the strips in the spring tends to maintain a thick grain brace.

Another option is to sow a good grain cover crop in the autumn. It is essential to till the crop field early in the spring so that residues from the cover crop do not conflict with the application of plastic mulch, drip irrigation and fumigation (if necessary). When the safety of the wind is no longer needed, the grain strips are mowed and used as drive lines for insect spraying and disease control, and later for harvesting.

Plastic flower pots and field harvesting containers

Plastic flower pots are now widely used for the producing flowers as plastic is lightweight and more durable, thus shipping costs are lower and thus injury losses are sustained. Clay is a lot heavier and requires a lot of volume for the same number of items. Plastics raw materials can raise the cost to a large degree, even by recycling. Plastic pots are going to increase the quality. There would be a rise in the price of clay pots, but mostly as a consequence of rising shipping prices. Manufacturing clay pots is a very ancient technique, but the method of fire can pose issues in terms of both expense and emissions. Plastic pots are only economical when manufactured in vast quantities, causing expenditure in heavy machinery. As far as aesthetics and purpose are concerned, without touching the objects, it is difficult to say the difference between a terracotta clay and a plastic bowl, at least in larger sizes and when fresh, so there is not a lot to choose between the two from a horticultural pov, so it's down to your own tastes.

Disposal of the used Plastics from Plasticulture

The elimination of the plastics used from plasticulture is definitely a worldwide epidemic. There have been several efforts to fix issues, including the production and usage of photodegradable or biodegradable fabrics (which only disappear), the use of plastics many times (delaying the problem), the reduction of film weight (less waste to be treated), recycling (making films into new products) and incineration. Plastic mulches and drip irrigation tapes are the worst to treat. Following a season in the field, these products are dusty and mostly muddy, rendering them impossible to recycle. One choice is to incinerate them in order to restore their very significant fuel worth. A pound of plastic contains as many BTUs (thermal units) as an equal volume of fuel oil.

Using plastics have been burning in waste-to - energy facilities, but one concern is that they create hot spots in the waste stream. Another choice is to produce a "fuel nugget" that can be used in addition to coal or other waste materials or used on its own to heat different forms of structures. Further work needs to be conducted on the correct processing, storage and transfer of plastics from the site of use to the stage that it is refined and eventually to the stage of incineration. Incineration might be the solution to the question of disposal, however far further research is required.

Current and Future Status of Biodegradable Plastics

Biodegradable polymers have recently been produced for agricultural use. Biodegradable films are frequently thinner than conventional polyethylene, although often they are very identical. Biodegradable films are released in white, black and a number of colours. They can be manufactured from recycled products such as starch, cellulose or degradable polymers. Two forms of biodegradable polymer films are currently being studied. One is a polyethylene with a patented additive that enables the oxidative degradation of the polyethylene material. The breakdown process is a two-step procedure. The first step is an oxidative degradation that breaks down the long carbon chains into smaller parts. The oxidative decay is induced by radiation, humidity and mechanical tension. The film is porous and the molecular components are further broken down through microbial activity on CO2, soil and natural substances.

The second stage is the operation of the microbial. Depending to how the film is made, the period of time it takes to fall down (from many months to multiple years) would be influenced. They are not the same as photodegradable mulches that were historically accessible leaving plastic residues in the ground.

The other forms of films under study are polyester dependent starch. Starch is applied to the solution to break down long polymer (carbon) chains to shorter shapes, which are then further broken down by microbial action into CO2 and liquids. Polyester is biodegradable, too. Polyethylene film with starch added to improve breakdown is also available, but the issue is that polyethylene molecular fragments can remain in the soil.

There are compostable standards in the USA (ASTM) and Europe (EN) which specify that the following conditions must be fulfilled in order for a commodity to be compostable:

1. **Disintegration,** the capacity to break into non-distinguishable sections and to safely promote bioassimilation and microbial development.

2. **Biodegradation,** oxidation of carbon to CO_2 at 60 % over 180 days (ASTM standard) and 90 % over 180 days (EN standard).

3. **Safety,** that there is no proof of eco-toxicity in the finished compost and soil may help plant production.

4. **Toxicity,** where the quantities of heavy metals are fewer than 50% of the acceptable levels. No goods actually meet with compostable requirements. However, the materials may be biodegradable, but in a longer timeline.

Biodegradable films are usually thinner and thus special caution must be taken when setting the film to avoid tars and punctures. Weed management must be successful under the mulch because weeds expand from biodegradable films instead of getting smothered because frequently happens under stronger polyethylene films.

Chapter - 12

Artificial Intelligence, IoT and Robotics for Hi-tech Horticulture

Introduction

The value of modern applied technology in the agricultural sector, in order to improve not only the production capacity and quality of agricultural goods, but also the energy output of greenhouses and the mitigation of emissions, has demanded new climate management systems. The structure of the greenhouses has therefore evolved from simple plastic covered greenhouses to computer controlled greenhouses, making progress in this respect. Automation in agriculture, called Precision Agriculture (PA), is one of the main concepts of this process. Over the last few years, the development of Internet sector has made in smart greenhouses where the automation of the climate environment allows visual information and tasks to be carried out through algorithms which include the prediction and classifying procedure based on the Artificial Intelligence (AI) models. Despite the fact that the climate is controlled in a smart greenhouse, there is still a problem of maintaining the most favourable climate for the proper development of plants, because it is highly dependent on the external climate. In fact, there are significant fluctuations in the temperature and humidity of the air within the greenhouse during the day.

What is Artificial Intelligence?

Artificial Intelligence (AI) is now a popular term, but it could be uncertain whether this technology will simplify activities and increase greenhouse production. AI is a lot better than you might imagine at first. AI is nothing more than a complex software program. Using artificial intelligence, Motor Leaf builds sophisticated software that is essentially 'a lot of information' by finding patterns and data associations. When the trends are observed by the AI system, you will provide the AI with fresh insights and say, "What do you suppose is going to happen under these conditions?" Then the AI platform can give a high degree of precision.

No need for robots, but just as smart

The kind of AI technology that we use to develop our greenhouse automation services is computer science. No engines apart from powerful computers are involved, so no, in this case there seem to be no robots designed to automate greenhouse tasks. As the term suggests, after-harvest cycle, our greenhouse technology is learning and being more and more reliable as we train technology with your increasing results.

For example, our greenhouse clients can obtain 50-70 per cent decrease in harvest prediction error from the beginning; after a year of further learning, AI technology helps to reduce errors by more than 70 per cent. The AI system would also be more efficient in the forecasting of harvest yield weeks through the future.

Agriculture IoT: En Route to Intelligent Farming

By 2022, the global IoT greenhouse demand will grow to $1.3 billion (up from $680 million in 2016). However, use of the Internet of Things in agriculture is not restricted to linked greenhouse solutions. Precision farming with GPS technology, drones that reduce replanting costs by up to 1000 per cent, automated machines, sensor-based livestock monitoring systems, etc. Agriculture firms are undertaking projects on the Internet of Things with the goal of automating time-consuming operations, minimizing the use of fertilizer, avoiding overwatering, reducing operating costs and improving total land productivity.

The Internet of Things (IoT) involves the development of artificial intelligence and smart technology for common household and industrial applications. This helps in the development of a creative community capable of making smart choices. Here we focus our attention on IoT-enabled greenhouses, which have the capacity to transform agriculture and improve the efficiency of this sector.

Working of a greenhouse

Commercial greenhouses typically sustain a steady high temperature and humidity standard. Control is provided by means of screens, temperature maintenance and it might be a computer that controls the overall activities of the environment. The

IoT-enabled greenhouse would definitely offer the highest results, as it can use autonomous data analysis to make critical decisions in a smart manner, without human interference.

>> **Automated, precise harvest forecasting using AI**

Using broad data from that conditions, AI is able to forecast the amount of future harvests with unprecedented precision. These findings show how a large-scale greenhouse-grown tomato on the vine (TOV) was able to automate harvest forecasting, a system that a team of highly skilled farm workers had previously undertaken a day-and-a-half of work per week. Greenhouse AI Software is getting stronger and better over time. It's a good play.

>> **Why does AI solve problems with harvest forecasts for greenhouses?**

There's many variables in your greenhouse that have an effect on plant growth and product maturation. It is impossible for people to analyze all these factors and understand precisely how their plants can grow. AI allows us to analyze all of these growth factors and offers very accurate assessments of plant growth. For this reason, AI works so well to estimate future harvests.

>> **AI can learn farming tasks and become more accurate with more data**

Human capacity to forecast future harvest outcomes frequently yields nuanced, imprecise performance. Once continually equipped with more than one data from each crop season, AI's automatic crop prediction can gradually become more reliable. AI is better than skilled agronomists in forecasting harvest yields, helping you to automate this task so that your head growers and agronomists can dedicate their energy to more important issues. On average, agronomists spend half a day free making harvest forecasts.

Smart Greenhouse Design

1. **Requirement analysis:** The standards shall be specified by the agronomist and the information and communication technicians. The agronomic approach introduces requirements for crops, soil, climate, water, nutrients, energy and irrigation. The ICT (Information and Communication Technology) strategy promotes the usage of emerging tools for development and optimisation. Through this research, agronomists and ICT technicians set out the specifications for the implementation of new technologies and skills through automated systems. The key components of the agricultural development cycle and their interaction

with ICT technologies are as follows:

» **Crop**: Temperature, sun, water, oxygen, mineral nutrients and other supports are the basic features. Sensors with that kind of data are used as inputs for automation and optimization operations. RGB sensors collect regular photographs that can be analyzed to examine the nature of development. User can design different types of processes using the image data.

» **Soil**: The roots of the plants are surrounding by nutrients that sustain the developing plant. Soil moisture, conductivity or disease control must be treated with sensors and actuators.

» **Climate**: Plants are growing within a narrow temperature range. Humidity and temperatures that are too high or too small limit efficiency. Vegetable plants and many flowers require a great deal of sunlight. These plants can be grown with special plant-growth lamps. Sensors and actuators are required in the same way as in previous instances.

» **Water and nutrients**: Water temperature, pH and electroconductivity (EC) are factors that need to be regulated. Soilless growth requires effective and successful solutions for hydroponic nutrients. Liquid nutrients (nitrogen, phosphorus, potassium) are produced by the agronomist.

» **Energy**: Monitoring energy usage and renewable generation are important requirements for the design of sustainable installations.

» **Irrigation**: Control of irrigation time is crucial. The plant continues to receive moisture through the irrigation process. Irrigation optimization strategies are critical for enhancing processes.

» **User interfaces, data collection, control and connectivity systems** need to be planned. Tables and graphs of graphical details are presented in real time. Core data is processed through IoT tools. Subsequent research provides knowledge on the growing method.

» **Event identification, classification and data analysis are key criteria** for the existing value-added services. Such facilities may be planned and built in either greenhouse subsystem. Additional management and development tools need to be added.

2. Smart greenhouse model: Gardware and software ecosystem

Automated greenhouse systems automatically monitor the atmosphere, field irrigation and vent opening behavior. These systems are configured by the technician manual system, introducing the setpoint values. The agronomist chooses the environment framework, the timetable for irrigation and the nutrient requirements for irrigation water. Many of these devices have a subsystem-based architecture (climate, drainage, automated vent openings). Such subsystems are built without the arrangement of interactions between them. Irrigation technology is not compatible with climatic conditions. The closure and raising of automated vent openings is therefore not connected to other subsystems. Another significant aspect is that the three subsystems do not have the capabilities to anticipate occurrences that will immediately alter the value of such variables when the setpoints are set. This is essential in the enhancement and improvement of this type of facility. The paradigm presented in this research, after studying various forms of automation in greenhouses, aims to turn such situations into intelligent systems, with the potential to anticipate automated behaviour and combine all subsystems to make them interoperable. The architecture of this application offers a solution that can already be found in automatic greenhouses. The concept suggests the following:

» Hardware interface that uses built-in tools to incorporate automatic subsystems for greenhouses.

» Software platform with artificial intelligence concepts performing forecast and classification techniques to automate new actions.

For these two systems, automated greenhouses are spaces for the deployment of technologies focused on modern smart technology. First, the layout of hardware that incorporates the different subsystems (climate, crop irrigation, vent opening and renewable generation) should be interoperable in the design and deployment of smart services. The architecture proposed is focused on two stages. A level created by programmable control devices attached to each already mounted subsystem. The controllers are linked to each other and to the local network, which is controlled by a gateway. Control nodes communicate with second-level network management servers using IoT protocols.

» **IoT communication:** When a node applies a forecasting or classification model, a multi-process operating system (Linux) is needed. Unless the node only implements a control thread, a micro - controller with a key characteristics thread is needed.

» **Processing capacity:** One of the most commonly employed IoT protocols is Message Queuing Telemetry Transport (MQTT). The control node must provide the communication interfaces required to transfer data using this form of communication protocol.

Iot Greenhouse: First Step Towards Smart Agriculture

From pesticides to surface temperatures and quality, the number of factors affecting plant growth is literally endless. Thus, a farmer may either hire an army of agricultural staff to track crops on a 24/7 basis or go out of business. Now there's another option: you can trust routine tasks including lighting control and plant growth tracking to smart sensors and apps with built-in analytics capabilities. The connected greenhouse is a farming facility that incorporates microcontrollers, sensors and software from the Internet of Things; it often works in conjunction with other technical solutions put in place in agriculture, including automatic irrigation and HVAC systems. Smart sensors collect data on plant production, irrigation, pest usage and lighting and send it to an on-site or cloud-based server; a network admin console allows farmers to customize and connect the device settings with other applications, while a smartphone application creates IoT greenhouse warnings and reports.

Creating a wireless IoT greenhouse (as opposed to a wired one) is the best way to turn to insight-based farming without significant capital investment; the expense of constructing and sustaining a grid of 1,000 miles of wiring would undoubtedly impact the prices of agricultural goods (decreasing because they are).

Fig 10.1: IoT based Greenhouse
Source: https://r-stylelab.com/

How to Build a Smart Greenhouse?

First and foremost, you can collaborate with a reputable IoT technology firm that has already been interested with Smart Farming ventures. Your consultant can then select the best application platform for the project depending on the scale of your property, the type of plant you produce and the IT technologies that you have already deployed.

Step 1: Choosing hardware based on project requirements

The IoT microcontroller unit (MCU) is at the center of every wired greenhouse device. Depending on the capability and processing resources, a microcontroller may use firmware or a classic operating system to process and pass sensor data to a server. If you do not choose to mount a handheld camera that allows real-time plant tracking, you should go for a device that embraces several types of sensors, uses Bluetooth communication, absorbs minimal power and can be installed into the soil or connected to the stems.

More advanced IoT greenhouse applications using cameras are operated by microcomputers such as BeagleBoard, Arduino and Raspberry PI. They use classic C++ built-in software to process sensor data and show other Machine Learning capabilities (these devices , for example, evaluate the amount of green or red in the image, approximate harvest maturity, and submit data to the central base statio).

As a rule, one microcontroller covers up to 30 meters of arable ground, so you only need five sensors to create a 1,000 m² greenhouse. The MCUs are then linked to the power grid. Although microcontrollers consume relatively little power (150 mA with both BLE and Wi-Fi data connections enabled and just 5 mA with deep sleep mode on), the introduction of an IoT greenhouse system would hardly impact your energy bill.

Step 2: Evaluating connectivity options

The next move is to facilitate connectivity between the microcontrollers that make up the IoT network. You should opt for a typical Bluetooth network system where nodes share data and re-transmit messages received by a remote MCU before the target is reached. In fact, the provider will migrate the network to a configurable platform that handles certain MCUs. You should spread a variety of Wi-Fi routers around the greenhouses, allowing sensors to connect to the Internet and transmit data back to the server. The option of a networking system depends on the form of sensors and MCUs you are installing, the frequency of the Wi-Fi signal and the total region of the arable field.

How much does it cost to design an IoT greenhouse?

Based on our expertise in the creation of the Industrial Internet of Things, we estimate the cost of a personalized IoT greenhouse system (minus hardware) at $100-150 thousand. The volume includes the development of integrated applications (LUA / C++), a web-based framework (MySQL and PHP stack) and a smartphone alarm software, as well as advisory services on the option of hardware components.

Advantages of IOT based greenhouse

The usage of IoT in greenhouses offers various benefits. Here are some of the top advantages of using the IoT-enabled greenhouse for agricultural needs:

Minimizing waster use

A main concern in agriculture is the calculation of the exact amount of water needed. Minimizing the usage of water is feasible by adding smart sensors powered by the IoT program. This facilitates the setting up of water policies where the water tap can be regulated dynamically with the use of strategically located proximity sensors.

» **Plant conditioning**

The optimal plant cooling is another benefit available in the IoT powered greenhouse. It is done through the use of cameras and other sensors that can track the state of the plants on a daily basis and also automatically send a warning when a issue can be detected. Using the perfect cloud-based IoT system, this data can be collected and checked regularly to insure that all plants obtain the optimal treatment in the greenhouse.

» **Ideal atmosphere**

There are many factors that converge to create the optimal environment for a particular plant. Temperature, airflow, oxygen and carbon dioxide must be controlled under tight limits. It is achievable in an IoT powered greenhouse, where smart sensors will exchange their knowledge with each other, allowing improved decision-making.

Atmospheric environments may be set up in cloud-based IoT applications as a series of guidelines. It can then be combined with the IoT powered sensors mounted in the greenhouse. The machine will use the available readings to make decisions on running displays, use fans and other tools to retain power.

» One of the best ways to build an IoT enabled greenhouse is to install advanced, smart sensors which can be managed and controlled by IoT applications such as IoT Sense. It allows us to look at the accessible data analytics and set up automatic action policies to achieve the best agricultural conditions.

» This method aims to track and regulate the climatic factors that are conducive to the growing of a given plant. By utilizing this method, crop growth can be increased along with maximized production, irrespective of weather conditions.

» This project could be further expanded to track and regulate the pesticide amount.

Smart Agriculture applications account for 6% of all IoT programs. The global agricultural sensor built at the base would hit 600 million units in just four years. Through 2050, the Internet of Things technology should improve food production through 70%. In order to minimize running costs and make the most of the accessible arable property, farmers are willing to hop into the IoT bandwagon, and their path most frequently begins with the creation of a wired greenhouse.

Robots in Greenhouse Horticulture

1. Spraying robots the most popular

The most widely used robots in greenhouse horticulture are robot sprayers (24.7%). In addition to spraying machines, robots (22.2 %) are used for planting and harvesting. Packaging robots (11.7%) and piling robots (3.7%) are also included.

2. Staff shortage

Within the same survey, 23 per cent of greenhouse growers say that they are more likely to experience workforce shortages. The problem is that robotization is the solution to the lack of workers.

Mechanization and invention actually need less workers and, in addition, their number in agriculture has been decreasing for years. This is not always the solution, though, as growing fruit and vegetables is less easy to mechanize than, for example, arable goods.

3. Long term solution

Robotisation is also not the remedy for short-term workforce shortages: "While horticulture is revolutionary, such as the installation of robots harvesting ripe fruits with the aid of image processing, implementations are still fairly small and the costs are significant. Plant breeding may make it possible to optimize the harvesting cycle in the long run. Through this view, robotization just provides a long-term response to the lack of workers in the industry."

Chapter - 13

Integrated Nutrient Management for Hi-tech Horticulture

MINERAL NUTRITION

It has been known for centuries that the roots of terrestrial plants have been nourished from the soil. Within the first half of the 19th century, it was observed that plants required some chemical elements referred to as essential elements and that these elements were consumed primarily by the roots as inorganic ions. These inorganic ions in soils are often derived from soil mineral constituents. The term mineral nutrient is generally used to refer to the inorganic ion produced from the soil and required for plant growth. The process of absorption, translocation and assimilation of plant nutrients is known as mineral nutrition.

Essential Elements

Plants require 17 elements for their development and the completion of their life cycle. They are: carbon, hydrogen, oxygen, nitrogen, phosphorus, potassium, calcium, magnesium, sulfur, iron, manganese, zinc, copper, boron, molybdenum, chlorine and nickel. In addition, four extra elements, i.e. sodium, cobalt, vanadium and silicon, are processed by some plants for special purposes. Not all of these elements are needed for all plants, but all of them have been found to be necessary for one plant or another. Among these, all carbon atoms and most oxygen atoms are derived from carbon dioxide, which is assimilated mainly in photosynthesis. More specifically, nearly one - third of the oxygen atoms in the organic compounds in higher plants are extracted from soil water and two thirds from atmospheric carbon dioxide. The chemical basis for this conclusion is the reaction that occurs during photosynthesis, in which one molecule of carbon dioxide (CO_2) and water (H_2O) is integrated in the presence of enzymes. Items C, H, O are not minerals. The rest of the elements are absorbed from the soil and are called mineral elements because they are derived from minerals. These mineral elements are primarily absorbed in ionic form and, to some extent, in non-ionic form as shown in Table 13.1.

Criteria of Essentiality

Plant research using modern techniques reveals that the plant body comprises about 30 elements and, in certain instances, up to 60 elements. The presence of several elements in the plant does not mean that all of them are essential for plants. Arnon and Stout (1939) suggested essentiality parameters to be refined by Arnon (1954). An element is considered necessary when plants cannot complete the vegetative or reproductive stage of the life cycle due to its deficiency; when this deficiency can be repaired or avoided only by providing this element; and when the element is specifically involved in the metabolism of the plant.

Table 13.1: Forms of mineral elements absorbed by plants

Nutrient	Chemical Symbol	Principal forms for uptake
Carbon	C	CO_2
Hydrogen	H	H_2O
Oxygen	O	H_2O, O_2
Nitrogen	N	NH^{+4}, NO^{-3}
Phosphorus	P	H_2PO^{-4}, HPO_2^{-4}
Potassium	K	K^+
Calcium	Ca	Ca^{2+}
Magnesium	Mg	Mg^{2+}
Sulfur	S	SO_2^{-4}, SO_2
Iron	Fe	Fe^{2+}, Fe^{3+}
Manganese	Mn	Mn^{2+}
Boron	B	H_3BO_3
Zinc	Zn	Zn^{2+}
Copper	Cu	Cu^{2+}
Molybdenum	Mo	MoO_2^{-4}

Chlorine	Cl	Cl⁻
Nickel	Ni	Ni^{+2}

At a realistic point of view, certain parameters are deemed too strict. According to these criteria, sodium is considered to be non-essential. However, sodium is known to increase the yield of a number of crops, such as sugar beets, turnips and celery. As a consequence, the farmer finds sodium to be an integral factor. Nicholas (1961) proposed the term 'functional nutrient' for any mineral element that functions in the metabolism of plants, whether or not its action is specific. With these criteria, sodium, cobalt, vanadium and silicon are also considered functional nutrients. In addition to 17 primary components.

Classification of Essential Elements

Essential elements may be categorized on the basis of the quantity needed, their mobility in the plant and soil, their chemical composition and their role within the plant.

> **Quantity of nutrients**

Based on the amount of nutrients present in plants, they could be grouped into three: basic nutrients, macronutrients and micronutrients.

> **Basic nutrients:**

Carbon, hydrogen and oxygen, which make up 96 per cent of the overall plant dry matter, are basic nutrients. Carbon and oxygen make up 45 per cent of each.

> **Macronutrients:**

Nutrients required in large amounts are known as macronutrients. They're N, P, K, Ca, Mg and S. N, P and K are considered primary nutrients, while Ca, Mg and S are classified as secondary nutrients. The latter are known as secondary nutrients because they are inadvertently applied to soils by means of N, P and K fertilizers containing these nutrients.

> **Micronutrients:**

Nutrients needed in small amounts are known as micronutrients or trace elements. It's Fe, Zn, Cu, B, Mo, Cl and Ni. Such components are very effective and achieve optimal results in small quantities. On the other hand, even a slight deficiency or excess is harmful to plants.

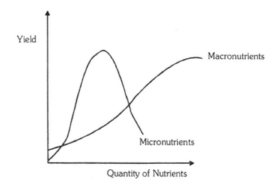

Fig. 13.1: Yield Curves for micro and macro nutrients

Source: http://www.agricultureinindia.net

Functions in the Plant

Based on the functions, the nutrients are grouped into four components: basic structure, energy use, charge balance and enzyme activity.

1. Elements that provide the basic structure of the plant-C, H and O.

2. Useful elements for energy storage, transfer and bonding-N, S and P. These are structural accessories that are more active and vital to living tissues.

3. Elements necessary for the balance of charges-K, Ca and Mg. They serve as regulators and carriers.

4. Elements involved in enzyme activation and transfer of electrons-Fe, Mn, Zn, Cu, B, Mo and Cl. These elements are catalysts and activators.

Mobility in the Soil

The persistence of nutrients in soils has a significant effect on the supply of nutrients to plants and the process of fertilization. Two mechanisms are essential for plants to pick up such nutrients: (1) the transfer of nutrient ions to the absorbing root surface and (2) roots to the region where nutrients are available. In the case of immobile nutrients, the roots must meet the supply of nutrients and the thickness of the forage is restricted to the root level. The whole volume of the root zone is the forage area for highly mobile nutrients (Fig. 13.2).

Dependent on the movement of the surface, the nutrient ions can be classified as mobile, less mobile and immobile. Mobile nutrients are highly soluble and are not adsorbed to clay complexes, e.g. NO^{3-}, SO_4^{-2}, BO_3^{-2}, Cl^-, Mn^{+2}. Less mobile nutrients are also soluble, but they are adsorbed to the clay complex and thus limit their mobility, e.g. NH^{4+}, K^+, Ca^+, Mg^{+2}, Cu^{+2}. Immobile nutrient ions are extremely reactive and are trapped in soil, e.g. H_2PO^{4+}, HPO_4^{-2} Zn^{+2}.

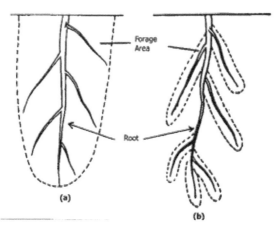

Fig. 13.2. Forage area for (a) mobile and (b) immobile nutrients
Source: Reddy and Reddy (2016)

Mobility in Plants

Awareness of the availability of nutrients in a plant allows to determine whether a nutrient is deficient. In case of deficiency, the mobile nutrient in the plant goes to the growing stage. The symptoms of deficiency therefore appear on the lower leaves.

1. N, P and K are very mobile.

2. Zn is moderately movable.

3. S, Fe, Mn, Cu, Mo and Cl are less mobile.

4. Ca and B are immobile.

Chemical nature

Nutrients may be categorized into cations and anions and metals and non-metals on the basis of their chemical composition.

Cations: K, Ca, Mg, Fe, Mn, Zn, Cu

Anions: NO_3, H_2, PO_4, SO_4

Metals: K, Ca, Mg, Fe, Mn, Zn, Cu

Non-metals: N, P, S, B, Mo, Cl

Nutrient Availability

Factors influencing nutrient availability

Several factors influence the availability of nutrients:

1. The natural supply of soil nutrients closely linked to the parent material of the soil and the vegetation under which it is developed;

2. Soil pH, since it influences the release of nutrients,

3. Relative activity of microorganisms which play a key role in the release and fixation of nutrients;

4. Fertility in the form of agricultural fertilizer, livestock manure and natural manure,

5. Temperature , humidity and aeration of the soil and

6. Presence of plants and type of cropping system.

The presence of plants increases the availability of nutrients in soils, particularly in the rhizosphere. Fungal species and phosphatase processes are found in the plant rhizosphere. While the rhizosphere is colonized by 23 fungi organisms, it is dominated by Penicillium species. Plant age has an influencing impact on the amount of microorganisms and the production of phosphatase. The impact becomes more prominent throughout the flowering season. There is a positive correlation between the occurrence of microorganisms and the activity of phosphatase, indicating that rhizosphere-associated mycoflora plays a significant role in the transformation of phosphorus.

Successfully growing plants interact with the chemical transformation of P in soil by complex processes (hydroxylation of soil colloids, chelation of metal cations; P diffusion in soil) that are powered by chemical energy from photosynthetic products and result in plant-induced fertilization of soil. As a consequence of these methods, a certain volume of soil and nutrient P is removed and ultimately taken up by the plants.

Transport of Nutrients to Root Surface

Two key theories, namely soil solution theory and contact exchange theory, explain plant availability of nutrients.

>   **Soil solution theory.** Soil nutrients are absorbed in water and transferred to root surfaces through mass flow and diffusion. Mass flow is the movement of nutrient ions and salts with moving water. As a consequence, the movement of nutrients entering the root depending on the amount of water flow. Diffusion happens because there is a concentration difference between the root surface and the underlying soil solution. Ions move from the high concentration region to the low concentration region.

>   **Contact exchange theory.** Close contact between root surfaces and soil colloids allows for a direct exchange of H^+ from plant roots with soil colloids cations. The importance of the exchange of contact in the transport of nutrients is less important than the movement of soil solution.

Mechanism of absorption

Nutrients are absorbed by the plants in two ways:

>   **Active absorption.** Absorption of nutrients from soil water having a low concentration of nutrients relative to plant sap, through exhausting energy, is called active absorption.

>   **Passive absorption.** Nutrients reach plants in the transpiration path without the cost of energy.

Translocation

Nutrients absorbed by plant roots enter the cortex region where they are collected against the concentration gradient. They enter xylem vessels from the cortex and enter the leaves through mass movement along with the transpiration current. As far as nitrogen is concerned, the portion of absorbed nitrate-nitrogen (NO_3-N) is reduced to ammonical-nitrogen (NH_4-N) and to root glutamine. These compounds, along with the remaining portion of NO3-N, pass through the simplest (living link between cells) and enter the xylene. From the xylem vessels, the mass flow in the transpiration stream moves upwards. They end up in leaves where nitrate reduction occurs. Reduced compounds enter phloem vessels and are passed to growing areas such as young leaves, roots, fruits, etc.

Functions of Nutrients in the Plant

Basic nutrients, C, H and O, are carbohydrate components and several biochemical compounds. Nirtogen is a component of proteins, enzymes, hormones, vitamins, alkaloids, chlorophyll, etc. Plant growth is adversely affected by nitrogen deficiency as it is a component of enzymes, chlorophyll and proteins. Phosphorus is a component of sugar phosphates, nucleotides, nucleic acids, coenzymes and phospholipids. The cycle of anabolism and carbohydrate catabolism continues as the organic compounds are esterized with phosphoric acid. Potassium is not a component of any organic compound. However, 40 or more enzymes are required as a cofactor. Controls the movement of the stomata and maintains the electroneutrality of the plant cells. Sulfur is a component of several amino acids and fatty acids.

Calcium is a component of the cell wall as a calcium pecate. It is needed as a cofactor in ATP and phospholipide hydrolyses. Magnesium is a chlorophyll constituent and is necessary in many enzymes involved in the conversion of phosphate. As a constituent of various enzymes (cytochrome, catalase, dipeptides, etc.), iron plays a vital catalytic role in the plant. As a result, it is a key element in various redox respiration reactions, photosynthesis and nitrate and sulphate reductions. Manganese is a part of many cation activated enzymes such as decarboxylase, kinase, oxidase, etc., and is therefore necessary for the formation of chlorophyll, nitrate reduction and respiration. Copper and zinc are active in the activated enzymes of cation. Boron can carry carbohydrates. It is necessary for the germination of pollen, for the formation of flowers and fruits and for the absorption of cations. Molybdenum is necessary both for the assimilation of nitrates and for the fixation of atomspheric nitrogen. Chlorine is involved in the reaction of the production of oxygen.

Roles of Elements in Plant Growth

Carbon is "fixed" by photosynthesizing plants from CO_2 in the atmosphere. Carbon is an element of all organic compounds, including sugars, proteins and organic acids. Such compounds are used, among others, in structural components, enzymatic reactions and genetic material. The respiration process degrades organic compounds that provide energy for various plant processes.

The normal level of CO_2 in the atmosphere is approximately 350 ppm. Study on many crops has shown that if the atmospheric amount of CO_2 is raised to 800 to 1000 ppm, plant growth and yield would improve. CO_2 injection is a standard practice in the production of greenhouse vegetables in northern climates in winter. Large amounts of CO_2 will be retained in these ecosystems as the greenhouses are locked throughout the winter. If CO_2 injection is done, the grower can't add CO_2 while the ventilation system is off, and CO_2 can't be applied during the daytime, because that is when the CO_2 will be absorbed by the plant.

CO_2 levels can be boosted by burning natural gas or liquid propane in special CO_2 burners, or they can be added into the house as CO_2 gas. Burners require a way of spreading CO_2 to the property. Usually this is done by blowers. The CO_2 gas is pumped into tubing, typically polyethylene, with emitters spaced around the tube.

Injection of CO_2 will be most effective if the injection is done inside the canopy of the plant where CO_2 can easily enter the leaves of the plant. Growers use CO_2 injection will invest in a CO_2 control device such that the amount of CO_2 will not increase to an unsustainable point. Monitoring and control of CO_2 injections can be computerized and automated.

» **Hydrogen** is also a part of organic compounds where we consider C. Hydrogen ions are engaged in electrochemical reactions to sustain load balances across cell membranes.

» **Oxygen** is the third element in traditional organic molecules, such as basic sugars. O is important for many plant biochemical reactions.

» **Nitrogen** is a very critical component of plant growth and is present in many compounds. That involve chlorophyll (green dye in plants), amino acids, proteins, nucleic acids and organic acids.

» **Phosphorus** is used in a number of energy transfer compounds in plants. The role of P in nucleic acids, building blocks for genetic code material in plant cells, is very important for P.

» **Potassium** is used as an activator in many plant enzyme reactions. Another function of K in plants exists in specific leaf cells called guard cells located around the stomata. Guard cell turgor (or lack of turgor) controls the degree of stomata opening and therefore controls the exchange of gas and water vapor through the stomata. Turgor is largely controlled by the movement of K in and out of the guard cells.

» **Calcium** is needed for the formation of calcium pecate in the cell wall. In addition, Ca is used as a cofactor for other enzymatic reactions. Recently, it has been identified that Ca is involved in the intimate control of cell processes regulated by a molecule named calmodulin.

» **Magnesium** plays an essential function in plant cells because it occurs at the middle of the chlorophyll molecule. Some enzyme reactions involve Mg as a cofactor.

» **Sulphur** is a part of sulfur-containing amino acids, such as methionine. Sulphur is also found in the sulfhydryl community of other enzymes.

» **Iron** is used in biochemical reactions formed by chlorophyll and is part of one of the enzymes resulting in a reduction of nitrate nitrogen to ammonia nitrogen. Other enzyme systems, such as catalase and peroxidase, require Fe as well.

» **Manganese** functions in several enzymatic reactions involving the adenosine triphosphate (ATP) energy compound. Manganese frequently stimulates many enzymes and is active in photosynthesis processes in the electron transport system.

» **Zinc** is involved in the activation of many enzymes in the plant and a growth regulator is needed for the synthesis of indoleacetic acid.

» **Copper** is an element of many enzymes in plants and is a component of the protein in the photosynthesis electron transport system.

» **Boron** functions are not well understood in the plant. Boron tends to be essential for the normal growth of meristem in young sections, such as root tips.

» **Molybdenum** is a component of two enzymes involved in the metabolism of nitrogen. Nitrat reductase is the most important of these.

» **Chlorine** plays a potential function in photosynthesis and may serve as a counter ion in the K fluxes involved in the cell turgor.

» **Nickel** is a component of certain plant enzymes, in particular urease, which metabolizes urea nitrogen into usable ammonia in the plant. Without nickel, toxic levels of urea may accumulate in the tissue that forms necrotic legions on the tips of the leaf. In this case, nickel deficiency causes toxicity to the urea.

Deficiency Symptoms

When a sufficient quantity of nutrient is not present, plant growth is affected. Plants may not exhibit visible effects of a certain amount of nutritional content, but development is impaired and this condition is recognized as hidden hunger. When the nutrient level is still low, the plants exhibit characteristic signs of deficiency. Such signs, depending on the variety, have a general sequence. These are generally masked by diseases and other stress and therefore require careful and patient observation of more plants for typical symptoms. Symptoms of deficiency are visible in crops with larger leaves.

Nutrient Deficiencies

Nitrogen can be consumed by the plant in either nitrate (NO_3) or ammonium (NH^{4+}) types. The NO_3 type is typically the desired source of supply of most N to greenhouse crops. The form of NH_4 appears to be absorbed more easily than NO_3 at cool temperatures (less than 55 F). NH_4 uptake is highest at medium pH above equilibrium with lower uptake as the pH decreases. Uptake of NO_3 is highest at higher stages of acidic pH. The maximum N absorption of plants typically happens as all sources of N are available in the environment. The presence of NH_4 in the NO_3 media resulted in the highest growth rates in some tests.

The manner in which N is ingested has an impact on the pH media. When NH4 is consumed, the plant emits H+ ions to preserve the electrical equilibrium and thus the pH decreases. When NO3 is consumed, pH rises due to the existence of higher levels of OH-ions. Therefore, the uptake of N can explain some of the variations often seen in the pH of the growth media. Nitrogen is a very mobile factor of the plant and thus the signs of deficiency occur first on the lower leaves. Symptoms include the general yellowing (chlorosis) of the stems. On tomatoes, the petioles and leaf veins can have some red coloration. Unless the condition continues, the leaves may fall off the vine.

Common plant leaves have a dry weight value of between 2.0 percent and 5.0 percent N. Deficiencies of N are most often found where errors are made in the management of fertilizers which result in insufficient supply of N to crops. More frequently than not, there is a issue with excess N demand. Plants with excess N are typically moist and delicate with wider, darker-green leaves. Excess N (especially in warm and sunny seasonings) can lead to "bullish" tomato plants. These plants develop dense, leathery leaves that curl under a dramatic pattern of compact growth.

Phosphorus is consumed as either $H_2PO_4^{-1}$ or HPO_4^{-2} by an active energy-requiring cycle. P is very mobile in the plant. Deficiencies thus occur on the older leaves of the plant as P is translocated from these leaves to satisfy the needs of the new growth. P deficiency occurs as stunting and reddish coloring owing to decreased levels of anthocyanin pigments.

Deficient leaves would have just around 0.1 percent P of dry matter. The most recently matured leaves of most vegetables should produce 0.25 to 0.6 percent P on a dry weight basis. Excess P in the root zone is likely to result in reduced plant growth as a result of P delaying the uptake of Zn, Fe, and Cu.

P absorption can be decreased by high pH in the root media or by cold media temperatures. It is necessary to seek to sustain the pH of the hydroponic solution at 5.6 to 6.0 in order to support P uptake. Acidification can be accomplished with the usage of many acids such as sulphuric, nitric or phosphoric acids. It is necessary not to oversupply the peat media in the trough community. Maintain a pH of 5.5 to 6.5. Test the temperature of hydroponic solutions and peat bags or other solid media. Media temperatures should not fall below 60F for prolonged periods of time, particularly during seedling growth. Hot media decreases the absorption of P by plant roots.

Occasionally, thin petioles and midrib leafs on young leaves of full-grown plants exhibit slight purpling. That is also the case in late fall as temperatures are high. This is presumably not linked to P deficiency because it exists on young leaves. This is presumably also related to elevated levels of anthocyanin pigment released in leaves at cool temperatures. It does not create issues and would vanish in colder circumstances.

Potassium is consumed in significant amounts through an aggressive absorption cycle. When in the plant, K becomes very mobile and is easily transferred to young tissues.

Deficiency symptoms for K first appear as marginal flecking or mottling on lower leaves. Prolonged deficiency results in necrosis along the leaf margins and the plants can become mildly wilted. Deficient plant leaves normally produce less than 1.5 per cent of K. K deficiency contributes to a blotchy ripening of tomatoes where the plant does not contain a natural red color in the plant areas. Excess levels of K in the media, particularly hydroponics and rock fiber, may inhibit the absorption of other cations, such as Mg or Ca.

Sulfur is primarily consumed in the form of sulfate (SO_4). Sulfur is not very flexible in the plant, so deficiency usually starts with new development. The symptoms of deficiency are the general yellowing of the leaves. The deficiency of N and S is similar, but the deficiency of N occurs on the lower leaves; the deficiency of S occurs on the upper leaves.

Plant leaves typically have a dry weight value of between 0.2 percent and 0.5 percent S. This spectrum is close to that of P. Plants can generally tolerate quite high levels of S in growing media, and this is one reason for the widespread use of Sulphur-containing materials to supply nutrients such as Mg and micronutrients. S deficiency is not very common in greenhouse vegetable crops for this reason.

Calcium, unlike other components, is consumed and transferred through a passive process. The cycle of plant transpiration is a significant factor in the absorption of Ca. When in the plant, calcium travels into areas with a large degree of transpiration, such as quickly developing plants.

Most of the absorption of Ca happens in the root area just below the root tip. This is of vital interest to greenhouse vegetable growing, as it ensures that farmers will preserve stable root systems with plenty of active root tips. Root diseases will severely restrict the plant's calcium uptake.

Calcium becomes immobile in the plant, so the signs of deficiency occur first on fresh growth. Ca deficiency causes necrosis of new leaves or leads to curled, contorted growth. Sources of this are tip-burning lettuce and cola crops. Blossom-end-rot of tomato is also a calcium deficit associated condition. Tomato fruit cells deprived of Ca break down creating the well-recognized dark region of the tomato fruit blossom-end. Often this breakdown will occur just inside the fruit, such that tiny darkened hard patches develop within the tomato while the outside appears to be natural. Occasionally the lesion on the surface of the fruit may be sunk, or it can merely be a darkening of the tissue around the center of the flower.

As the movement of Ca in the plant is linked to transpiration, it follows that the environmental factors that cause transpiration often influence the movement of Ca. High humidity intervals can contribute to the burning of the tip of the lettuce, as the leaves may not transpire rapidly enough to transfer sufficient Ca to the growing leaf extremities.

Calcium concentrations in normal, most recently matured leaves will range from 1.0 percent to 5.0 percent. Deficiencies, however, may occur on a temporary basis under certain environmental conditions as discussed above. It is therefore important to consider the control of irrigation and the greenhouse environment in the overall Ca fertilization programme. In fact, the absorption of Ca can be impaired by other ions such as NH_4, Mg, and K. These cations can compete with Ca for root uptake. These conflicting resources will not be provided in excess of what the plant requires.

Magnesium is absorbed by the plant at a proportion greater than Ca. The absorption of Mg is often highly impaired by competing ions such as K, Ca, or NH_4. In comparison to Ca, Mg is elastic in the plant and the deficiencies occur first on the lower leaves.

Mg deficiency consists of interveinal chlorosis, which may contribute to necrosis in the infected regions. Advanced Mg deficit in tomato leaves contributes to moderate purpling of infected regions.

Mg is usually found in normal leaves at concentrations of 0.2 per cent to 0.8 per cent. Conditions that contribute to deficiency involve badly engineered fertilizer systems that provide too little Mg or those that provide excess K, Ca, or NH_4.

Iron may be extracted through an active mechanism such as Fe^{2+} or by iron chelates, which are organic molecules comprising iron sequestered within the atom. The uptake of iron is heavily dependent on the iron type and the correct uptake relies on the capacity of the root to reduce Fe^{3+} to Fe^{2+} for uptake. Iron chelates are soluble and tend to hold Fe in a solution for absorption. The uptake of the entire chelate molecule is low and Fe is usually removed from the chelate prior to uptake.

Iron is not elastic in plants, so the signs only occur on fresh seeds. Symptoms consist of interveinal chlorosis, which can contribute to bleaching and necrosis of the infected plants. Common leaves produce 80 ppm to 120 ppm Fe on a dry weight basis.

Conditions contributing to Fe deficiency include low amounts of Fe in nutritional water, cold air, or alkaline air (pH over 7.0). Fe deficiency is cured by applying Fe to the fertilizer solution or by spraying Fe foliar. Generally one or two sprays with a solution of 25 ppm Fe (chelated iron product) can fix a temporary deficiency of Fe.

Manganese is consumed as Mn2 + ions and is influenced by other cations such as Ca and Mg. Manganese becomes fairly immobile in the plant and the signs of deficiency occur on the upper leaves.

The deficit of Mn is close to that of Mg, but Mg occurs on the lower leaves of the herb. Mn deficiency consists of interveinal chlorosis; indeed, chlorosis is more evident than magnesium deficiency. Natural concentrations of Mn in leaves vary from 30 ppm to 125 ppm for most plants. Large amounts of Mn can be harmful to plants. Toxicity is a partial leaf necrosis in many species. Concentrations of Mn in the range of 800 ppm to 1000 ppm can contribute to toxicity in many crops. Excess

Mn decreases the absorption of Fe in the nutrient solution.

Situations that contribute to deficiency are often linked to insufficient availability of Mn in solution or to the competing impact of other ions. Toxicity can occur from excess Mn supply or from acidic media containing certain soil-based content that carries Mn. Solubility of Mn is improved by low pH in such soil-based media.

Zinc absorption is considered to be an active mechanism and can be influenced by the accumulation of P in the media. Zn is not very mobile in plants. Zn deficiency results in leaves with interveinal chlorosis. Often Zn deficiency can contribute to plants with reduced intestines.

Common leaves produce around 25 ppm to 50 ppm of zn. High concentrations of Zn may contribute to toxicity when root development is reduced and the leaves are small and chlorotic. Zinc deficiency can be enhanced by cool, wet rising water, or by means with a very high pH or an abnormal P.

Copper is consumed in very low amounts by plants. The absorption cycle tends to be an active mechanism that is heavily influenced by Zn that pH. Copper (Cu) is not highly mobile in plants, although certain Cu may be translocated from older to newer leaves. The normal level of Cu in plants is between 5 - 20 ppm.

Copper deficiency of young leaves causes chlorosis and some elongation of the leaves. Excess copper, particularly in acidic media, can be toxic.

Molybdenum is consumed as Molybdate MoO_4^{2-} and the absorption can be blocked by sulfate. Tissue quality of Mo is normally less than 1 ppm.

First, a shortage of Mo occurs in the center of the leaves and the older plants. The leaves are chlorotic and the margins drop. Like other micronutrients, the deficit of Mo exists mainly under acidic environments.

Boron: The absorption of boron by plants is not well known. Boron (B) is not elastic in the plant and tends to have certain characteristics of absorption and transport in comparison with Ca.

Boron deficiency first effects young development stages, e.g. leaves, leaf edges, and margins. Budds grow necrotic areas and the tips of the leaf become chlorotic and ultimately die. Tomato leaves and stems are becoming brittle. Average leaves produce 20 ppm to 40 ppm B although high levels can contribute to toxicity. Only limited quantities of B are required by plants and the oversupply of B from nutrient solutions or foliar sprays contributes to toxicity.

Chlorine deficiency is only occasionally found in crop plants. It is because Cl is required in very limited amounts, and Cl is found in fertilizers, water, air, and media in the atmosphere.

Nickel: In the past, nickel (Ni) was not known to be an important element for plant development, however work has established that it is an appropriate element for plant growth. The standard range for nickel in most plant tissues is about 0.05-5 ppm. Owing to its low requirements (often in parts per billion), it can be present at adequate amounts as a contaminant in dirt, water, fertilizer, etc. Nickel deficiency is rare and is sometimes misdiagnosed as there are originally no signs in plants. It illustrates why most labs will not check it and that it is not used in most fertilizers.

Small nickel deficiency has no visible effects, but may limit plant growth and yield. Significant nickel deficiency shows visible signs usually in the old leaves of the plants, as nickel is a mobile element. Deficiency signs of legumes are known as whole leaf chlorosis coupled with necrotic leaf tips (caused by the accumulation of toxic urea levels). In woody ornamentals, signs of new emerging growth arise in spring and can include shortened intestinals (giving a rosetary look to the plant), low shoot development, death of terminal buds, and ultimate death of shoots and branches.

Identification of Deficiency Symptoms

Symptoms of deficiency may be distinguished on the basis of (1) the region of occurrence (2) the presence or absence of dead spots and (3) the chlorosis of whole leaf or intervenal chlorosis. (Fig. 13.3)

The area of presence of symptoms of deficiency relies on the availability of nutrients in plants. Nutrient deficit effects in lower leaves owing to their versatility within plants. Such nutrients transfer from lower leaves to rising leaves, triggering signs of deficiency in lower leaves.

Zinc is moderately mobile in plants and therefore the symptoms of deficiency appear in the middle leaves. The signs of deiciency of fewer mobile elements (S, Fe, Mn and Cu) occur on fresh leaves. Since Ca and B are immobile in plants, signs of deficiency occur in terminal buds. Chlorine deficiency in crops is less normal.

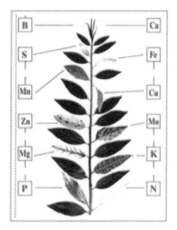

Fig. 13.3. Identification of deficiency symptoms
Source: *University of Arizona Cooperative Extension via P interest*

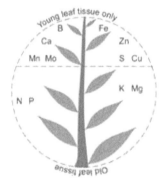

Fig. 13.4. Region of occurrence of deficiency symptoms
Source: https://grdc.com.au/

Deficiency Symptoms on Old Leaves

The signs that occur on old leaves can be further identified by the appearance or lack of dead spots.

Without dead spots. The typical deficiency signs of nitrogen are the continuous yellowing of the branches, including the roots. The leaves become rigid and erect, especially in cereals. After a little vigorous pull, the leaf can detach in severe deficiency in dicotyledonous crops. Cereal crops have a distinctive 'V' shaped yellowing at the tip of the lower leaves.

In phosphorous deficiency, the leaves are small, upright, rarely dark green with a greenish-red, greenish-brown or purplish tinge. The back side creates a bronze look.

Magnesium deficiency also causes yellowing, but differs from nitrogen deficiency. The yellowing takes place in the middle of the veins and the veins remain green. The leaf isn't upright. The leaf is splitting very quickly and can be removed by blowing the wind. For severe situations, necrosis (death of tissues) happens only in the edges.

With dead spots. In potassium deficiency, the yellowing begins with the tips or margins of the leaves extending to the center of the base of the leaf. These yellow parts will soon become necrotic (dead spots). There is a sharp difference between green and yellow and necrotic portions. Dead spots occur particularly on the margins and tips.

Molybdenum deficiency produces transparent patches of abnormal form between the branches of the trees. These spots are either green, yellow or gray. The infected areas are impregnated with resinous gum, which emanates from the back of the leaf from the reddish brown patches.

Deficiency Symptoms on New Leaves

These signs can extend through the whole leaf or the veins may remain green.

Veins remaining green. The veins remain green in the deficiency of iron and manganese. Under iron deficiency, the primary veins stay conspicuously green and the other parts of the leaf switch gray towards whiteness. Much of the leaf is under serious deficiency. It's gray. In manganese deficiency, the key veins as well as the smaller veins are black. The interveinal portion is yellowish, not white. Dead spots may also occur at a later level. There's a confirmed presence on the vine.

Veins not remaining green. The leaf is yellowish owing to a sulfur deficiency, but it behaves like a nitrogen deficiency crop. The leaf is thin and the roots are paler than the intervenial section. There are no empty places. The lower leaves of the plant are not damaged as in the case of N deficiency.

In the case of copper deficiency, the leaf is yellowish, leading towards whiteness. In extreme deficiency, vein chlorosis occurs and leaf loss of luster occurs. Leaf is unable to maintain its turgidity, and thus wilting happens. Leaf is removed from the base of the petiole owing to water saturated circumstances.

Terminal buds

The signs of deficiency of Ca and B are also seen on fresh plants. Nevertheless, it is easier to identify the signs of their failure on terminal buds or growing areas than on fresh leaves.

Through calcium deficiency, the bud leaf is gray with the base staying black. About one-third of the chlorotic part of the tip sticks down and becomes brittle. Loss of terminal buds happens in severe situations.

Boron deficiency causes yellowing or chlorosis, starting from base to tip. The edge is rather elongated like a whip like a structure and is brownish or blackish brown. Loss of terminal buds happens in severe situations.

Deficiency on Both Old and New Leaves

In the case of zinc deficiency, the leaf is narrow and tiny. Lamina is chlorotic, and the veins stay black. Subsequently, dead spots form in the vine, including roots, tips and margins. Zinc deficiency usually exists in cereals in 2-4 leaves from the top during the vegetative period. Plants look bushy due to decreased internal elongation. Subsequently, the panicle struggles to arise entirely or appears partly.

Sources of Nutrients

Materials used to provide nutrients for greenhouse vegetable growth are selected on the basis of many criteria, including cost per unit of nutrients, solubility in water, capacity to provide several nutrients, freedom from pollutants and ease of handling. Such products are often used for the preparation of a liquid fertilizer nutrient solution.

Premixed fertilizer products are very common with many greenhouse operators as they are simple to use. However, they are relatively expensive compared to individual ingredients, and the premixed materials leave little room for changes in the concentrations of individual nutrients. Pre-mixed fertilizers are most usually supplied with P, Mg, S, and micronutrients, and portion of N. We can even supply some or more of the K.

Pre-mixed fertilizer products are very familiar among many greenhouse operators as they are simple to use. However, these are fairly costly compared to raw components, so premixed products do not provide room for improvement in the ratios of specific nutrients. Premixed fertilizers are more usually provided with P, Mg, S, and micronutrients and a part of N. We could even supply some or more of the K.

Crop Nutrient Requirements

Efficient greenhouse vegetable crops need the necessary quantity of growing nutrient, the crop nutrient requirement (CNR) throughout the season. The growth rate of the plant determines, to a large extent, the amount of nutrients needed in the nutrient solution during the growth cycle. This plan is most suitable for crops grown in some forms of solid soil, such as peat, where the soil can contain considerable quantities of nutrients. Supplying plants with limited quantities of nutrients on a continuous basis should be as rewarding as practicable.

This scheduling strategy has not been carried out for greenhouse vegetables in all Florida development systems. Growers should bear in mind that these formulations are useful starting points and that they will need to be changed in any way, based on increasing experience and on unique site-specific analyzes of water. Water analyzes will most often affect the amount of Ca, S and acid added to the fertilizer. More information is provided in the section on pH control.

Water Quality for Greenhouse Vegetables

In the production systems of NFT, water is the means of production in which the plants grow. Water makes up a large portion of the growth media for most solid-media cultural systems. It is also quite appropriate that the consistency of the output material, water, should be identified. This is somewhat similar to soil checking for nutrient-supply potential in field scenarios. Without an accurate water test, a grower cannot claim to have a good fertilizer management programme.

Water is the means of production in the production systems of NFT in which the plants grow. Water is a large part of the growth media for most solid-media cultural systems. It is also quite appropriate to identify the consistency of the output material, water. This is somewhat similar to soil monitoring of the potential for nutrient supply in field scenarios. Without an accurate water test, a grower cannot claim to have a good fertilizer management program.

Water quality is determined by the quantity of particulate matter (sand, limestone, organic matter, etc.), the quantity of dissolved materials (nutrients and non-nutritional chemicals) and the pH of water. These aspects must be determined by a good water analysis, such as that conducted by the Florida Extension Soil Testing Lab or by a competent commercial laboratory.

In certain Florida cases, the laboratory will track electrical conductivity (EC), hydrogen ion concentration (pH), sulfate concentration (SO_4), water hardness, sodium (Na) concentration, chloride (Cl) concentration, iron (Fe) concentration and bicarbonate (HCO_3) concentration. Hardness is relative to the Ca and Mg concentrations in the water and determination of these two nutrients will help the grower in calculating the Ca and Mg fertilizer program.

Water quality is determined by the quantity of particulate matter (sand, limestone, organic matter, etc.), the quantity of dissolved materials (nutrients and non-nutritional chemicals) and the pH of water. These aspects must be determined by a good water analysis, such as that conducted by the Florida Extension Soil Testing Lab or by a competent commercial laboratory.

In certain Florida instances, the laboratory will test electrical conductivity (EC), hydrogen ion concentration (pH), sulfate concentration (SO4), water hardness, sodium (Na) concentration, chloride (Cl) concentration, iron (Fe) concentration and bicarbonate (HCO_3) concentration. Calcium concentrations of 50 ppm Ca or higher are very normal in Florida well water. The concentrations of magnesium are not as high, usually less than a few ppm. Very high concentrations of Ca and Mg and high bicarbonates could lead to precipitation of calcium and magnesium carbonates and clogging of the emitters of the irrigation system.

The concentrations of iron in some Florida wells are relatively high (0.5 ppm or more). Concentrations higher than 0.5 ppm could lead to precipitation of iron resulting in the plugging of irrigation emitters.

In fact, large concentrations of S in well water are not dangerous from a plant nutrition point of view. Nevertheless, a high concentration of S will contribute to the build-up of sulphur-bacteria in the irrigation lines that could obstruct the emitter.

Bicarbonate (HCO_3^{-1}) concentrations are frequently in excess of acceptable amounts. Large concentrations (above 30 ppm to 60 ppm) are correlated with decreased pH values. Over time, high HCO_3^{-1} rates will contribute to a rise in pH values in the growth media. High HCO_3^{-1} will also contribute to precipitation of Ca and Mg carbonates. Owing to possible complications associated with high levels of HCO_3^{-1} it is advised that water with high levels of HCO_3^{-1} be acidified at pH 5.6 to 6.0 with nitric, phosphoric or sulphuric acids.

Media Reaction, pH

The pH of the media refers to the concentration of hydrogen ions (H^+) in the media solution. The concentration is measured by a pH electrode or may be approximated by a pH color strip of paper dipped into the solution. The pH of the media solution is important because some aspects of plant nutrition are influenced by pH, such as the solubility of essential elements. Most of the elements are best absorbed from a pH media of 5.5 to 6.5.

Media pH above 7.0 results in reduced micronutrient and phosphorous solubility. Extremely acidic pH can lead to micronutrient toxicity, especially in soil-based media, if manganese and aluminum are present.

Media pH values may vary over time depending on the uptake of certain macronutrients. For example, nitrate uptake (NO_3^{-1}) can lead to an increase in the pH solution. This is because the plant is trying to maintain an electrical load balance across the membranes; therefore, the hydroxyl (OH^-) ion is exuded from NO_3 uptake.

Absorption of K+ has the opposite effect as H+ is exchanged resulting in the acidification of the media. This rapid change in media pH is most prevalent in hydroponics and rockwool compared to solid media culture because there is no buffering capacity in the former. Nutrient solutions must either be supplemented on a regular basis or acids or bases must be applied to maintain the correct pH levels.

Most well water in Florida has a fairly high pH content, typically attributed to high levels of bicarbonate (HCO_3^{-1}). Water analyzes should determine the levels of bicarbonate so that the amounts of acid to be used to reduce the pH can be determined. More information is provided in the Water Quality section.

Foliar Sprays

Many farmers are concerned about the effectiveness of foliar sprays of nutrients or other additives, such as antiperspirants. In general, these products have not been identified as having a place in a well-managed greenhouse.

Fertilizer plans are structured to provide the plant with nutrients by root systems. The roots are very well able to consume nutrients, and the leaves are not. Leaves can consume a small amount of nutrients, but the absorption of other components is insufficient. The leaf is covered with a waxy layer, a cuticle, and this structure renders it impossible for nutrients and other contaminants to reach the leaf in vast amounts.

Often, special miracle products are advertised by salesmen for use to prevent "stress" and that routine sprays will improve yield and quality. Plants under threat are more prone to have temperature or light issues rather than nutrient issues. In many field vegetable research studies, it has been shown that routine foliar spray shotguns do not increase yield or improve quality. In reality, several studies indicate a decrease in yield from these foliar sprays. There is a big difference between shotgun spray and foliar nutrient spray to treat a specific deficiency. Iron deficiency is one of the areas where a foliar nutrient spray may be recommended. Often the upper leaves of a tomato plant that turn yellow, particularly after a large number of tomatoes have been planted. A foliar solution spray of 0.5 to 1.0 ml of a 5 percent Fe solution in 1 gallon of water will usually clear up the problem. Repeated sprays may be needed. The foliar spraying of micronutrients must be carried out on a diagnostic basis and care must be taken to ensure that the correct amount is applied. Foliar burns can quickly occur if caution is not taken.

Many foliar sprays, such as growth regulators and anti-transpirants, have not reliably shown good results in field crops. Good plants that usually work are unlikely to react to a miracle drug. In the case of antiperspirants, it is unlikely that they will have the desired effect in the greenhouse. In most poly-covered greenhouses, desiccation not a problem because it is usually very moist inside and water loss is not a big problem. In addition, in greenhouse situations where CO_2 is injected, antiperspirants may have a negative effect.

In summary, typical plants are unlikely to react to extra stimuli from foliar chemicals. Growers could use shotgun foliar products to waste money and even damage their crops.

Organic manures used under Hi-tech horticulture

Organic products have a longer residual impact in addition to enhancing soil physical, biological and chemical properties. The practice of processing and utilizing compost from animal, human crop and vegetable waste to increase crop production is as ancient as agriculture. Significant origins of organic materials are as follows:

1. **Cattle shed wastes:** Dung, urine, and slurry from biogas plants

2. **Human habitation wastes:** night soil, human urine

3. **Poultry litter:** Dropping of sheep and goat.

4. **Slaughter house waste:** bone/ meat meal, blood meal, horn/hoof meal

5. **Fish wastes**

6. **By-products of agro-industries:** oil cakes, bagasses and pressmud, fruit and vegetables, processing wastes etc.

7. **Crop wastes:** sugarcane trash, stubbles and other related material.

8. **Water hyacinth:** weeds and tank silt

9. **Green manure crops and green leaf manuring material.**

Farm Yard Manure (FYM)

It is a decomposed combination of dung and urine of farm animals, along with waste feed, litter, etc. On average, well-decomposed farm yard manure produces 0.5 percent N, 0.2 percent P_2O_5 and 0.5 percent K_2O. Trench system of preparation of FYM, the usage of chemical preservatives such as gypsum, gober gas plant slurry will greatly minimize the storing and handling of nutrient losses to increase the consistency of FYM. The trenches are 6 m to 7 m in length. 1.5 m to 2.0 m in width and 1.0 m in depth was drilled. Any available litter and waste is combined with compost and applied to the shed to absorb urine. The next morning, the urine, soaked in the waste along with the dung, is gathered and deposited in the trench. For filling with regular set, a portion of the trench from one end should be taken up. When the portion is filled up to a height of 45 cm to 60 cm above the ground floor, the top of the heap is turned into a dome and covered with cow dung earth slurry. The phase begins and the second trench is prepared until the first trench is entirely completed. The manures are ready for usage in around four to five months after the plastering phase.

Compost

The method of decomposition of organic residue is called composting and the decomposed substance is called compost. Composting is basically a microbiological decomposition of organic residues collected from rural areas (rural compost) or urban centers.

Farm compost is created by putting farm waste in trenches of the required scale. Say 4.5 to 5 m long and 1.5 to 2.0 m wide and 1.0 to 2.0 m deep. Farm waste is deposited layer by layer in the trenches. Each layer is well moistened with the sprinkling of cow dung slurry or water. The trenches are packed to a height of 0.5 m above the ground. The compost is ready for usage within five or six months. Composting is performed in either aerobic or anaerobic environments. Any approaches are susceptible to these factors. The drawback of the aerobic method is that it is quick but needs moisture and regular rotating.

Poultry manure: Bird excreta degrade very easily. If left exposed, 50% of the nitrogen would be lost within 30 days. Poultry manure contains higher amounts of nitrogen and phosphorus relative to other bulky organic manures. The mean nutrient value is 3.03% N 2.63% P_2O_5 and 1.4% K_2O.

Oil-cakes: After the oil is removed from the oil seeds, the resulting solid part is dried into a layer that can be used as a manure. There are two types of oil-cakes:

1. **Edible oil-cakes** that can be comfortably fed to animals, e.g. groundnut cake, coconut cake, etc.

2. **Non-edible oil-cakes** that are not ideal for feeding animals, e.g. castor cake, neem cake, mahua cake, etc.

Both edible and non-edible oil cakes may be used as manures. However, edible oil cakes are fed to cattle, and non-edible oil cakes are used as manures, especially for horticultural crops. Nutrients present in oil-cakes, after mineralization, shall be made available to crops 7 to 10 days after application. Until use, oil-cakes must be well powdered for even distribution and faster decomposition.

Bone meal: Bones from slaughtered animal carcasses, slaughterhouses and the beef processing industry are specific suppliers of bone meal. Bones are a rich supply of phosphorus and calcium. Crushed bones are used and in raw form or steam sterilization. Glue extracted from bones has a market benefit and residues in powder form are used either as compost or as feed for livestock. Bone meal is slow-acting and suitable for dry soils and long-lasting crops.

Guano: It is a mixture of excreta and dead bodies of sea birds abundant in N and P. It is obtained from islands on a regular basis. Since removing oil from fish in factories, the refuse stayed dry and used as manure. Identified as a guano shark. It is comparable to the guano bird in its impact on soil and crops.

Crop residues: Residues left over after processing economic parts are considered crop residues / straw. In developing countries like India, we are mainly used as feed for livestock. In developed countries, harvesting is carried out utilizing the field itself. Straw has good manor worth because it provides an appreciable quantity of plant nutrients. On average, cereal straw and residues comprise approximately 0.5 % N, 0.6 % P_2O_5 and 1.5 % K_2O. Crop residues may be recovered through incorporation, composting or mulch material.

Table 13.2: Nutrient content in different organic manures

Manure	Nutrient (%)		
	Nitrogen (N)	Phosphorus (P₂O₅)	Potash (K₂O)
1. Bulky Organic Manure			
Farm Yard Maure	0.5-1.5	0.4-0.8	0.5-1.9
Compost (Urban)	1.0-2.0	1.0	1.5
Compost(Rural)	0.4-0.8	0.3-0.6	0.7-1.0
Green manure (averages)	0.5-0.7	0.1-0.2	0.6-0.8
Sewage sludge dry	2.0-3.5	1.0-5.0	0.2-0.5
Sewage allivated dry	4.0-7.0	2.1-4.2	0.5-0.7
2. Non-edible cakes			
Castor cake	5.5-5.8	1.8-1.9	1.0-1.1
Mahua cake	2.5-2.6	0.9-1.0	1.8-1.9
Karanj cake	3.9-4.0	0.9-1.0	1.3-1.4
Neem cake	5.2-5.3	1.0-1.1	1.4-1.5
Safflower cake (undecorticated)	4.8-4.9	1.4-1.5	1.2-1.3
3. Edible cakes			
Cotton seed cake (decorticated)	6.4-6.5	2.8-2.9	2.1-2.2
Cotton seed cake (undecorticated)	3.9-4.0	1.8-1.9	1.6-1.7
Groundnut cake	7.0-7.2	1.5-1.6	1.3-1.4
4. Manure of animal origin			
Fish manure	4.0-10.0	3.0-9.0	0.3-1.5
Bird guano	7.0-8.0	11.0-14.0	2.0-3.0
Bone meal (row)	3.0-4.0	20.0-25.0	-
Bone meal (Steamed)	1.0-2.0	25.0-30.0	-
5. Straw and Stalks			
Pearl millet	0.65	0.75	2.50
Sorghum	0.40	0.23	2.17
Maize	0.42	1.57	1.65
Paddy	0.36	0.08	0.71
Wheat	0.53	0.10	1.10
Sugarcane trash	0.35	0.10	0.60
Cotton	0.44	0.10	0.66

Recycling of Farm Wastes

Farm waste may now be regarded to include animal waste such as cow and buffalo dung and urine, other animal and human excreta, cereal crop waste, pulse and oilseed, corn stalk, cotton, tobacco, sugar cane trash and agro-industry items such as oil-cakes, paddy husk and bran, bagasse, press mud, fruit and vegetable waste, etc.

Estimates of availability of agricultural waste indicate that the average value of crop waste is 350 mt and that of animal waste is 650 mt. Therefore, about 1000 mt of agricultural waste is available in the country. The disposal of farm waste provides the much-needed organic and mineral content to the soil. Because most recyclable waste is organic, it also contributes organic matter and plant nutrients to it. Nowadays, more effective and flexible recycling processes and technologies have become accessible that, if implemented to the appropriate size, can carry recyclable waste into the main stream of farm input management strategies.

Benefits of proper recycling

1. Supply of essential plant nutrients

2. Improving physical, chemical and biological properties of soil;

3. Reducing their aggregation at or close production sites;

4. Reducing health threats,

5. Enhancing the quality of the environment, and the quality of human life, not just water fish and plants,

6. Providing jobs and profits for many,

7. It shows in a strong and solid way that man is not just a waste producer, but also a wise use / manager.

Importance of recycling

Substantial plant nutrient quality is derived from soil through crop harvesting every year. In India, for example, harvested crops extract 9-10 million tons more N, P_2O_5 and K_2O per year than overall fertilizer adds. The balance of nutrients must also be reached from organic sources. In this sense, the processing of organic waste by consumers is important for the recovery of at least some of the plant nutrients obtained from soil for the conservation of humus and soil productivity.

Methods of recycling: Farm waste can be recycled in soil through a number of methods:

1. Incorporation (ploughed in the soil)

2. Burning

3. Surface mulching

4. Composting

1. **Incorporation:** Incorporation of maize, rice, sorghum and wheat straw in soil during the preceding Kharif season has had a beneficial impact on soil physical, chemical and biological properties. Farm waste of all kinds may be ploughed in the soil (0-20 cm layer). The introduction of mechanized farming in many advanced countries has culminated in a significant volume of straw / stalk crop being left in the field after harvesting. About 50 percent of the residues remain in the field of mechanized farming using a mix. After the harvesting of cotton, sugarcane, sorghum, etc., the usage of rotator immediately contributes small pieces of crop residues to the soil.

2. **Burning:** Significant amounts of cotton, pigeon pea, castor stalk and cane trash are available and are presently being burning in the field by several farmers. These are not recommended activities. Burning increases the losses of N, C, S and probably certain other nutrients due to volatilization and results in adverse soil conditions. Burning extracts Ca, Mg, and K from Residues, however raises future losses related to leaching and erosion.

3. **Surface mulching:** In fact, this is a special and easy approach for recycling. While mulch decomposition is a slow method, the biomass and C / N ratio of mulch is greatly reduced during the one-crop season, making it simpler to introduce mulch at a low cost throughout the next season. Experiments have shown that straw mulch has a beneficial impact on moisture retention, erosion control, weed prevention, temperature variations and a beneficial soil micro flora population that eventually results in improved crop yields.

4. **Composting:** Compost is a stable and sanitized composting system that is helpful to plant development. It is projected that between 7.1, 3.0 and 7.6 million tons of N, P_2O_5 and K_2O can be supplied with the usable organic waste in India. This organics must then be processed and placed to good use. In the context of these evidence, more focus is given to the advancement of composting technologies.

Table 13.3: Nutrient content in different crop residues

Sr. No.	Crop residue	Nutrient (%)			
		N	P_2O_5	K_2O	Total
1.	Rice	0.61	0.18	1.38	2.17
2.	Wheat	0.48	0.16	1.18	1.82
3.	Sorghum	0.52	0.213	1.34	2.09
4.	Maize	0.52	0.18	1.135	2.05
5.	Pearlmillet	0.45	0.16	1.14	1.75
6.	Barley	0.52	0.18	1.30	2.00
7.	Fingermillet	1.00	0.20	1.00	2.20
8.	Sugarcane	0.40	0.18	1.28	1.86
9.	Lantana	2.50	0.25	1.40	4.15

Principles of Composting

Compost making includes three important and vital scientific principles

1. Reduction of carbon: ratio of nitrogen to a satisfactory level (10:1 or 12:1)

2. Complete elimination of dangerous pathogens and weed seeds assured by high temperatures during decomposition and stabilization.

3. At an ideal temperature of 60-65⁰C, all dangerous contaminants must be decomposed.

Essential requirements for composting

1. Bulky organic waste such as stubbles, cotton stalks, tur stalks, groundnut shells, weed leaves, dust bin refuse, etc.

2. Suitable starter: cattle dung, urine, night soil, sewage, urea, rock phosphate or some other readily accessible nitrogenous material. The mixture can contain 1-1.25 percent N on a dry weight basis (i.e., C: N-30:1). Microbial communities have also accelerated decomposition.

3. Apply sufficiently water to maintain the substance moisture content at a level of about 50%.

4. Presence of adequate air supply, particularly in the initial stages of decomposition.

Method of Composting

While the usage of crop waste in crop production is documented from the earliest periods, organized composting practice was only started at the beginning of this century. In India, Howard and Wad (1931) did some pioneering research at Indore and Howler (1933) in Bangalore. Composting is performed either in an aerobic or anaerobic environment. Any approaches are susceptible to these factors. The drawback of the aerobic method is that it is quick but needs moisture and constant rotation.

Table 13.4: Potential of organic and biological resources in India

Sr. No.	Name of Resource	Annual production of biomass (mt)	Nutrient supply (mt)
1.	Cattle and buffaloes (wet dung & urine)	2028.0	6.96
2.	Crop residues	336.0	8.74
3.	Forest litter	100.0	-
4.	City refuse	14.0	0.294
5.	Sewage sludge	6.0	0.011
6.	Press mud	5.0	0.266

Recycling of Wastes

Composting is an ancient process by which farmers have transformed herb, animal and human waste into organic manure, that is to say, added value to organic waste. Basically, any device or arrangement that ensures an effective decomposition of organic matter could be a composting process.

1. **Indore method:** Around 1930, Sir Albert Howard, a British agronomist in India, researched composting at Indore in a practical way and developed a modern composting system known as the Indore method. The waste materials are well mixed and well moistened with dung or nocturnal soil slurry and build into heaps of 4 to 6 m in length, 1 m in width and 1 m in height or into a pit of 30' x 5'x 3' with sloping sides. In the later method of charging a 30 ft pit in parts of 5 ft. the first segment is empty to allow mixing. The aerobic conditions in this system are preserved by occasional manual turning of composting materials into heaps and mixing products. If required, water is applied. In this aerobic phase, organic matter and nitrogen losses are in the region of 40-50 per cent of the initial amounts. Average manure composition was determined to be 0.8 % N, 0.3 % P_2O_5 and 1.5 % K_2O.

Fowler invented the 'activated compost' method in which fresh materials were inserted into the already fermented heap such that the already formed broad microbial community could contribute to faster decomposition. This method is especially useful when aggressive materials such as night soil need to be disposed of rapidly and efficiently.

2. **Bangalore method:** Dr. C. N. Acharya in 1938 developed a method for anaerobic composting of city garbage and night soil in pits. The trenches of following dimensions are dug in rows, roads of suitable width are provided between row for the carts to approach and unload the materials inside the trenches.

Table 13.5: Dimension of the trenches

Population ('000)	Length (m)	Width (m)	Depth (m)
<10	4.5	1.5	1.0
10 to 20	6.0	2.0	1.0
20 to 50	9.0	2.0	1.0
>50	10.0	2.5	1.0

The refuse and the night soil shall be spread in alternating layers of 15 cm and 5 cm until the pit is filled 15 cm above ground level with a fine layer of refuse at the top. It can be provided the form of a dome and covered with a thin layer of soil. Decomposition is mainly anaerobic but in the surface layer which is very slow. The C: N ratio is decreased to less than 20:1 in approximately six months. This process is also regarded as the hot fermentation method, since the loss of heat during decomposition is greatly decreased. While originally built for villages, it can be used to produce compost from readily accessible organic materials. Under rural settings, dung may be used to cover the night soil.

3. **Coimbatore method:** Waste composting is performed in the pits. The depth of 1.0 m and the width of 1.25 m allow fast handling when filling and turning the stuff. The duration of the pit depends based on the consistency of the material used for composting. A layer of waste material is laid out in the room. It is moistened with a slurry of 5-10 kg of cow dung in 2-2,5 liters. Water in which 0.5 to 1.0 kg of fine bone meal is applied. Related materials are placed one over the other before the substance reaches 0.75 m above ground level. It is plastered with wet mud and kept undisturbed for 8-10 weeks.

The mud plaster is discarded two months later when the substance is moistened, turned and formed into a rectangular heap in a shaded area. It remains undisturbed until it is required. At the process, an aerobic fermentation happens while the substance is kept covered with mud plaster. Aeration and aerobic fermentation occurs as formed in an open heap later. The tricalcium phosphate in the bone meal is soluble in the acids formed during decomposition and the compost is enriched by the addition of phosphorus. Compost is ready until the temperature in the pile is similar to that of the ambient environment. The finished result is black in color, fairly divided, high in humus and has a C : N ratio of 10:1 to 20:1.

4. **NADEP method:** This system of composting was created by the farmer Narayan Dev Rao Panthary Pande of Pusad Village in Maharashtra for the composting of farm waste. A concrete building with 22 cm thick walls and a height of 3.3 m L x 2.0 m W x 1.0 m H is built in a large and open environment with the aid of cement. For good ventilation, 10 cm x 10 cm holes are created in the wall, leaving the first and last line. This arrangement is called "TANKA" and can be used for longer periods of time. It is not necessary to fill the tank in one day; thus, the following materials should be obtained within 2-3 days.

Cow dung 60-100 kg; crop residue, green leaves, stalks, kitchen and feed waste etc. 1400-1500 kg; dry sieved soil 1700-1800 kg and water 1500-2000 lit. While filling the tank, the interior of the tank is sprayed with diluted cow dung slurry and then 15-20 cm thick layer of garbage (approximately 200 kg) is added to the bottom. The second layer is made of cow dung or biogas slurry by dissolving 4-5 kg of cow dung in 150 liters of water. The third layer of dry sieved soil (200-250 kg) free of stone, glasses and plastic, etc., is formed and the water is sprayed to moisten it. This step is replicated 7-8 times before the substance reaches 35-50 cm above the walls. It is then granted the form of a hut to confirm the entry of rainwater and plastered with slurry of soil and cow dung. After 45 days, the substance will be compacted down 25-30 cm and fractures will be evident on the soil. Now, the tanka is again full and plastered in the manner mentioned above. In this way, 3.5 t of high quality compost can be stored for 90-110 days. This approach is extremely useful when the supply of cow dung is limited. It is basically an aerobic process in which composting is conducted in a special perforated brick system to enhance aeration and reduce nutrient loses.

Bio Fertilizers for Organic Farming

Biofertilizers are known as preparations containing living cells or latent cells of productive strains of microorganisms that allow plants to absorb nutrients by interacting with the rhizosphere when applied via seed or soil. They promote some microbial processes in the soil, which increase the supply of nutrients in a manner readily assimilated by plants.

Microorganisms are most much not as effective in the natural world as one would assume, and thus artificially amplified colonies of efficient chosen microorganisms play a crucial role in speeding microbial processes in the soil.

The usage of biofertilizers is one of the main components of integrated nutrient management, as they are a cost-effective and organic source of plant nutrients to complement chemical fertilizers for sustainable agriculture. In the processing of biofertilizers, many microorganisms and their interaction with crop plants are being used. They may be classified in various forms on the basis of their existence and purpose.

Table 13.6: Classification of biofertilizers

Sr. No.	Groups	Examples
N$_2$ Fixing Biofertilizers		
1.	Free- living	Azotobactor, Beijerinkia, Clostridium, Klebsiella, Anabaena, Nostoc
2.	Symbiotic	Rhizobium, Frankia, Anabaena azolla
3.	Associative Symbiotic	Azospirillum
P Solubilizing Biofertilizers		
1.	Bacteria	*Bacillus megaterium var. phosphaticum, B. subtilis, B. circulans, pseudomonas striata*
2.	Fungi	*Penicillium sp, Aspergillus awamori*
P Mobilizing Biofertilizers		
1.	*Arbuscular mycorrhiza*	*Glomus sp., Gigaspora sp., Acaulospora sp., Scutellospora sp. & Sclerocystis sp.*
2.	*Ectomycorrhiza*	*Laccaria sp., Pisolithus sp., Boletus sp., Amanita sp.*
3.	*Ericoid mycorrhizae*	*Pezizella ericae*
4.	*Orchid mycorrhiza*	*Rhizoctonia solani*
Biofertilizers for Micro nutrients		
1.	Silicate and Zinc solubilizers	*Bacillus sp.*
Plant growth promoting Rhizobacteria		
1.	Pseudomonas	*Pseudomonas fluorescens*

Nitrogen fixing biofertilizers

(a) Symbiotic nitrogen fixers

1. *Rhizobium:* Rhizobium inoculants are commonly used by farmers around the world among all bio-fertilizers. These species colonize the roots of leguminous plants in order to form root nodules, which function as nitrogen input plants for the host plant. They live in these nodules and take air nitrogen and turn it into an organic form that the plant may use. As the bacteria live right in the root, they directly transfer nutrients to the plants. Rhizobium-leguminous symbioses can fix 100-300 kg N / ha on a yearly basis, depending on the crop, and live a considerable N content in the surrounding rhizosphere for the next harvest.

In legume, nitrogen fixation begins around 15 days after sowing when the nodules are tiny yet pink and occur at the time of flowering and early seed forming. It is therefore advisable to provide a starter dose of 20-25 kg N / ha to legume crops. *Rhizobium* will satisfy more than 80 per cent of the N needs of legume crops with 10-25 per cent rise in yield. Answer differs based on soil conditions and the productivity of the native community.

2. Azolla: *Azolla* is a floating water fern that is ubiquitous in distribution. It has an algal symbiont, *Anabaena azollae*, within its central cavity. Algae removes ambient nitrogen and is available at all levels of fern growth and production. *Azolla* comprises 0.2-0.3 percent N on a fresh weight basis and 3-5 percent N on a dry weight basis. It is used as a bio-fertilizer for rice in many countries and is comparatively more favorable than urea. It has the ability to repair more than 10 kg N / ha / day under optimal conditions. One crop of Azolla produces an improvement of 20-40 kg N / ha in around 20-25 days. During rice cultivation, farmers may take two such crops. Technology of *A. pinnata* production is known and well domesticated, which paddy growers may easily follow. *Azolla* fern has a greater capacity as a single crop in particular areas, such as low-lying fields, water-logged wastelands, seeding water, shallow wetlands, natural freshwater lagoons, burrow pits, Khet talawadi, etc., and thus generates jobs in rural areas. Over everything, *Azolla* is generally known as fertilizer, feed, food and fodder.

(b) Non Symbiotic nitrogen fixers

1. *Azotobacter*: These are free-living gram-negative rod-shaped nitrogen-fixing bacteria and are loosely associated with plants. They are natural soil inhabitants, ubiquitous in geographical distribution, and may have a variety of C and energy for their development. These bacteria will substitute 20-40 kg of N / ha for different crops. In India, based on soil fertility, the most probable number (MPN) of N fixators is slime, which aids in soil aggregation. In seeds, seed germination and plant stand are enhanced upon inoculation with better strains. Various species of *Azotobacter* are *A. agilis, A. chrococcum, A. beijerinckii, A.vinelandi, A. ingrinis*. A. Chrococcum is the primary inhabitant of arable soils and is most successful and commonly utilized. It improves crop yield by 10-15% and aids in the mineralization of plant nutrients and the dissemination of other beneficial micro-organisms.

2. *Azospirillum:* *Azospirillum* is a non-symbiotic N fixing bacteria. This is a very effective and commonly used biofertilizer for many crops in today's agriculture. This bacteria are closely connected with the roots of cereals and grasses. The individual cells are gram-negative curved rods, 1 mm in diameter, size and shape varying. There are four different species; *A. lipoferrum, A. brazillense, A. amazonense* and *A. halpraeferans*. The mechanism from which inoculated plants derive positive benefits is the same as *Azotobacter* and correct 20-40 kg N / ha in field conditions with an improvement in yield of 10-15%.

3. *Acetobacter:* It is a sacharophilic bacteria that links to sugarcane, sweet potatoes and sweet sorghum plants and binds 30 kg N / ha / year. This bacteria is usually leads for the sugar cane crop. It is known to increase the yield by 10-20 t / ha and the sugar content by about 10-15%.

Phosphate solubilizing biofertilizers

PSMs comprise different groups of microorganisms, such as bacteria, yeasts and fungi, which convert insoluble inorganic phosphates into soluble forms. Common genera such as *Bacillus, Pseudomonas, Aspergillus, Penicillium, Fusarium, Micrococcus* etc. have been reported to be active in PO_4 bioconversion. It is estimated that in most tropical soils, only 25 per cent is available for plant growth and 75 per cent of the superphosphate used is fixed. Important species of PSM includes *Bacillus polymaxa, B. coagulans, B. circulans, Psuedomonas striata, Aspergillus awamori* and *Penicillum digitatum*. PSMs can be multiplied by mass on Pikovasky's broth and mixed with carrier material for field use. These organisms possess the ability to induce phosphate solubilization by secreting organic acids such as formic, acetic acids, propionic, lactic, glycolic, succinic acids, etc. These acids lower the pH and cause bound phosphate to dissolve. The integrated use of PSMs could bring benefits from the low-grade rock phosphates available in our country to the tune of 230 million tonnes. PSMs are recommended for all crops in India and have been shown to replace 20-50 kg P_2O_5 / ha in different crops due to inoculative applications.

Liquid biofertilizers – A new panorama

Successfully produced liquid formulations of bio-fertilizers based on native culture Azotobactor and Phosphate, the substance has a minimum cell count of 10^9/ml and a shelf life of 1 year, ideal for drip irrigation and green house cultivation compared to currently market carrier-based (lignite) products with a shelf life of 6 months.

Dosage of liquid Bio-fertilizers in different crops

Table 13.7: Recommended liquid Bio-fertilizer and its method of use, the amount to be used for various crops are as follows:

Crop	Recommended Bio-fetillizer	Application method	Quantity to be used
Agro-Forestry/Fruit plants All fruit/ agro-forestry (herb, shrubs, annuals and perennial) plants for fuel wood fodder, fruits, gum, spice, leaves, flowers, nuts and seed purpose	*Azotobacter*	Soil treatment	2-3 ml/plant at nursery
Other Misc. Plantation Crops Tobacco	*Azotobactor*	Seedling treatment	1250 ml/acre
Tea, Coffee	*Azotobacter*	Soil treatment	400 ml/acre
Rubber, Coconuts	*Azotobactor*	Soil treatment	2-3 ml/plant
Leguminous plants/trees	*Rhizobium*	Soil treatment	1-2 ml/plant
Field crops, pulses, Chickpea, Pea, Groundnut, Soybean, Beans, Lentil, Lucern, Berseem, Green gram, Black gram, Cowpea and pigeon pea	*Rhizobium*	Seed treatment	500 ml/acre
Cereals, Wheat, Oat, Barley	*Azotobactor/ Azospirillum*	Seed treatment	500 ml/acre
Rice	*Azospirillum*	Seed treatment	500 ml/acre
Oil seeds, Mustard, Seasum, Linseeds, Sunflower, Castor	*Azotobacter*	Seed treatment	500 ml/acre
Millets, Pear millets, Finger millets, Kodo millet	*Azotobacter*	Seed treatment	500 ml/acre
Maize and Sorghum	*Azospirillum*	Seed treatment	500 ml/acre
Forage crops and Grasses Bermuda grass, Sudan grass, Napier Grass, Para Grass, Star Grass etc.	*Azotobactor*	Seed treatment	500 ml/acre

Note: Recommended doses when the inoculum count is 1×10^8 cells / ml, so doses would be 10 times higher than the above listed nitrogen fixings, Posphate solubilizers and potash mobilizers at a rate of 200 ml / acre for all crops.

Application of Bio-fertilizers

1. Seed treatment or seed inoculation

2. Seedling root dip

3. Main field application

Seed treatment: One packet of inoculant is combined with 500 ml of kanji rice to produce a slurry. The seeds needed for ha are mixed in the slurry in such a way that the inoculants are evenly coated over the seeds and then dried in the shade for 30 minutes. Shade dry seeds should be planted within 24 hours. One packet of inoculants (200 g) is adequate to handle 10 kg of seed.

Seedling root dip: This approach is used in transplanted crops. Five packets of inoculants are combined in 100 litres of water. The root portion of the seedlings needed for one ha shall be dipped in the mixture for 5 to 10 minutes and then transplanted.

Main field application: Ten packets of inoculants are combined with 50 kg of dried and powdered farm yard manure and then spread in one hectare of main field just before transplantation.

Rhizobium: *Rhizobium* is used as seed inoculants for all legumes.

Azospirillum/Azotobacter: In transplanted crops, *Azospirillum* is inoculated by seedling, root dip and soil application processes. *Azospirillum* is used for direct seed crops by seed treatment and soil application.

Phosphobacteria: Inoculated by seed, seedling root dip and soil application, as in the case of *Azospirillum*.

Combined application of bacterial biofertilizers: Phosphobic bacteria may be combined with *Azospirillum* and *Rhizobium*. Inoculants should be combined in equal quantity and added as stated above.

Points to remember

» Bacterial inoculants should not be combined with insecticide, fungicide, herbicide or fertiliser.

» Treatment of seeds with bacterial inoculants should be achieved at last while seeds are treated with fungicides.

Phosphobacteria: The recommended dose of Azospirillum is used for inoculation of phosphor-bacteria; for combined inoculation, both bio-fertilizers as recommended should be combined equally before use.

Vermicompost

Vermicompost is a type of composting by the use of earthworms, which typically live in soil and consume bio-mass and excrete it in a digested form. This compost is commonly referred to as vermicompost or vermicast, supplying essential macro elements such as nitrogen (0.74 %) P2O5 (0.97 %), K2O (0.45 %) and Ca, Mg and microelements such as Fe, Mo, Zn, Cu etc.

Benefits of vermicompost

1. When applied to clay soil, it loosens the soil to allow the air to penetrate.

2. The mucus associated with the casting is hygroscopic, collects water and avoids water retention and increases water holding capacity.

3. In vermicomposting, some of the secretions of worms and associated microbes serve as growth promoters along with other nutrients.

4. It increases the physical, chemical and biological properties of soil on repetitive applications over the long term.

5. Organic carbon in vermicompost releases the nutrient steadily and constantly into the environment and allows the plant to consume these nutrients.

6. The multi-faceted effects of vermicompost affect the growth and yield of crops.

7. Earthworm can mitigate the risks of contamination caused by organic waste by increasing the degradation of waste.

Methods of vermicompost

In general, the following are the three methods of vermicomposting under field conditions.

1. Vermicompost of wastes in field pits

2. Vermicompost of wastes in ground heap

3. Vermicompost of wastes in large structures

Vermicomposting of organic wastes in field:

Pits: It is desirable to use the optimal size of the ground pits and 10 x 1.0 x 0.5 m (L x W x D) will be the effective size of each vermicomposting site. The sequence of such beds is to be arranged in one place.

Ground heaps: Instead of opening pits, vermicomposting can be taken up in heaps of soil. The dome shaped beds (with organic waste) are packed and the vermicomposting is taken up. The optimum size of the ground heaps may be 5.0 x 1.0 x 1.0 m (L x W x H) heaps.

Composting in large structures: vermicomposting is taken up in large structures such as series of rectangular brick columns, cement tanks, stone block etc. which are filled with organic wastes and composting is taken up.

Each of these methods has got advantage as well as limitations. For example in (1) and (3) these would not be any mixing of soil with vermicompost unlike pit system, less incidence of natural enemies. But they need frequent watering (more of labour) compared to pit system. Similarly in places water is scarce (less rainfall tracts); pit system is good which in high rainfall areas (2) and (3) are advantageous as there would be proper drainage.

Steps: This is irrespective of methods

» **Selection of site:** it should be preferably black soil or other areas with less of termite and red ant activity, pH should be between 6 to 8.

» **Collection of wastes and sorting:** for field composting, raw materials are needs in large quantities. The waste available should be sorted in to degradable and non-degradable (be rejected).

» **Pre-treatment of waste:** Lignin rich residues – chopping and subjecting to lignin degrading fungi and later to vermibeds.

» **Crop stalks and stubbles:** dumping it in layers sandwiched with garden soil followed with watering for 10 days to make the material soft and acceptable to worm.

 » **Agro-industrial wastes:** mixing with animal dung in 3:1 proportion and later subjecting it for vermicomposting.

 » **Insecticidal treatment to site:** treating the area as well as beds (in case of pit system) with chlorpyriphos 20 EC @ 3.0 ml/liters to reduce the problem of ants, termites and ground beetles.

 » **Filling of beds with organic wastes:** wastes are to filled in the manner given below and each layer should be made wet while filling and continuously watered for next 10 days. In heaping and composting in special structures, the waste is to be dumped serially as done in pits.

Table 13.8: Filling of beds with organic wastes

7th Layer	A thick layer of mulch with cereal straw	(Top of bed)
6th Layer	A layer of fine soil (Black/garden soil)	(Top of bed)
5th Layer	Dung/FYM/Biogas sludge	(Top of bed)
4th Layer	Green succulent leafy material	(Top of bed)
3rd Layer	Dry crop residues	(Top of bed)
2nd Layer	Dung/FYM/Biogas spent sludge	(Top of bed)
1st Layer	Coconut coir waste/ sugarcane trash	(Bottom of bed)

Except 3rd and 4th layer (which is the material to be degraded) each layer should be 3 to 4 inch thick so that the bed material is raised above the ground level. Sufficient quantity of dry and green wastes is to be used in the beds.

 » **Introduction of worms in to beds:** the optimum number of worms to be introduced is 100 No/length of the bed. The species of earthworms that are being used currently for compost production worldwide are *Eisena foetida, Eudirlus eugeniae, Periony excavates, Lumbricus rubella* etc.

 » **Provision of optimum bed moisture and temperature:** Bed moisture: by watering at regular intervals to maintain moisture of 60 to 80% till harvest of compost. Temperature requirement for optimal results is 20-30°C by thatching (during summer). Monitoring for activity of natural enemies and earthworm and management of enemies with botanicals. Promising products: leaf dust of neem, *Acorus calamus* rhizome dust, neem cake etc.

Harvesting of vermicompost and storage: Around 90 days after release of worms, the beds would be ready for harvest. Stop watering 7days prior to harvest so that worms settle at bottom layer. Collect the compost, shade dry for 12 hours and bag it in fertilizer bags for storage.

Harvest of worm biomass: the worms are to be collected and used for subsequent vermicomposting.

Vermicomposting technique

» **Sheds:** For a vermicomposting unit, whether small or big could be of thatched roof supported by bamboo rafter and purling, wooden trees and stone pillars.

» **Vermi beds:** prepare 90 cm width, 45 cm height and length as per availability of dung and organic waste.

» **Land:** About 0.5 to 1 acre of land will be needed to set up a vermi compost unit cum extension centre.

» **Seed stock:** worms @350 per M^3 of bed space should be adequate to start with and build up the required population in about 2 to 3 cycles.

» **Water supply system:** To maintain optimum moisture content (40%) in vermibed, spray &apply water on vermin bed. Frequency & quality is regulated by prevailing climatic conditions.

» **Collection of VC:** When vermin compost is ready for collection, top layers apex somewhat dark granular and it used dry tea leaves have been spread over the layer. Watering should then stopped for 2-3 days and ready compost should be scrapped form top layers or to a depth.

» **Storage:** It should be stocked separately in bags. Before packing it should be sieved out from 2 cm galvanized mesh. Compost should not be exposed to sun.

Table 13.9: Average nutrient content of vermicompost

S. No.	Nutrient	Content	S. No.	Nutrient	Content
1.	Organic matter	30 to 40%	6.	Fe	120000 to 125000 ppm
2.	Nitrogen	1.50 to 2.0 %	7.	Zn	100 to 150 ppm
3.	Phosphorus	2.0 to 2.50%	8.	Mn	200 to 250 ppm
4.	Potash	0.6 to 0.80%	9.	Cu	20 to 30 ppm
5.	Ca	150 to 160 ppm			

Green Manure

Green undecomposed waste used as compost is considered green manure. It is produced in two ways: by growing green manure crops or by collecting green leaf (along with twigs) from plants grown in wastelands, fields and forests. Green manure is grown in field plants which typically belong to the leguminous family and are introduced into the soil after adequate growth to increase the physical structure as well as soil fertility. Plants grown for green manure known as green manure crops.

Methods of Green Manure

The practice of green manure is practiced in a number of forms in various countries or India to suit soil and climatic conditions. The techniques of green manure

1. Green manuring in situ

2. Green leaf manuring.

Green manuring in situ

Green manure crops are cultivated and introduced into the same field where they are cultivated, either as a pure crop or as an intercrop with the main crop, called green manure in situ. The most effective green manure crops are sunnhemp, dhaincha, cowpea, cluster bean and *Sesbania rostrata*.

Table 13.10: Biomass production and N accumulation of green manure crops

Crop	Age (Days)	Green biomass (t/ha)	N accumulated (kg/ha)
Sesbania aculeata (Dhaincha)	60	23.2	133
Sesbania juncea (Sunnhemp)	60	30.6	134
Vigna unguiculata (Cow pea)	60	23.2	74
Cymopsis tetragonaloba (Cluster bean)	50	20.0	91
Sesbania rostrata	50	25.0	96
Pillipesara	60	25.0	102

Table 13.11: Nutrient content of green manure crops

Plant	Scientific name	Nutrient content (%) on air dry basis		
		N	P_2O_5	K
Sunhemp	*Crotalaria juncea*	2.30	0.50	1.80
Dhaincha	*Sesbania aculeata*	3.50	0.60	1.20
Sesbania (Shevri)	*Sesbania speciosa*	2.71	0.53	2.21

Sesbania rostrata is a stem nodules a green manure crop native to West Africa. Since the plant is short-day and sensitive to photoperiod, the length of the vegetative period is short when planted in August or September. A mutant (TSR-1) developed by the Bhabha Atomic Research Centre, Bombay is insensitive to photoperiod, salinity-tolerant and waterlogged. Growth and nitrogen fixation with TSR-1 is higher than with current strains.

Green leaf manure

The use of green leaves and twigs of trees, shrubs and herbs collected from elsewhere is known as green leaf manure. Forest tree leaves are the main sources of green leaf manure. Plants growing in wastelands, field bunds, etc., are another source of green leaf manure. An important plant species useful for green leaf manure are neem, mahua, wild indigo, Glyricidia, Karanji *(Pongamia glabra)* calotropis, avise *(Sesbania grandiflora)*, subabul and other shrubs.

Table 13.12: Nutrient content of green leaf manure

Plant	Scientific name	Nutrient content (%) on air dry basis		
		N	P_2O_5	K
Gliricidia	*Gliricidia sepium*	2.76	0.28	4.60
Pongania	*Pongamia glabra*	3.31	0.44	2.39
Neem	*Azadirachta indica*	2.83	0.28	0.35
Gulmohur	*Delonix regia*	2.76	0.46	0.50
Peltophorum	*Peltophorum ferrugenum*	2.63	0.37	0.50
Weeds				
Parthenium	*Parthenium hysterophorus*	2.68	0.68	1.45
Water hyacinth	*Eichhornia crassipes*	3.01	0.90	0.15

Trianthema	*Trianthema portulacastrum*	2.64	0.43	1.30
Ipomoea	*Ipomoea*	2.01	0.33	0.40
Calotrophis	*Calotropis gigantea*	2.06	0.54	0.31
Cassia	*Cassia fistula*	1.60	0.24	1.20

Table 13.13: Potential of green manure crops

Green manure crops	Sowing time	Seed rate (kg/ha)	Biomass production (t/ha)	N (kg/ha)
Berseem (Trifolium alexandrium)	Oct–Dec	80–100	20–22	67–70
Black gram (*Vigna mungo*)	June–July	20–22	08–10	38–48
Cluster bean (*Cyamopsis tetragonaloba*)	April–July	20–22	10–12	40–49
Cowpea (*Vigna anguiculata*)	April–July	45–55	15–18	74–88
Daincha (*Sesbania aculeate*)	April–July	80–100	20–25	84–105
Green gram (*Vigna radiate*)	June–July	30–40	20–25	68–85
Horse gram (*Dolichos biflorus*)	June–July	25–30	26–30	120–135
Pea (*Pisum sativum*)	Oct–Dec	10–12	8–10	26–33
Sunhemp (*Crotolaria juncea*)	April-July	80-100	15–25	60–100

Characteristics desirable in legume green manure crops

> » Multi-purpose usage

> » Short duration, fast growing, high ability to absorb nutrients

> » Tolerance to shade; floods, droughts and adverse temperatures

> » Wider environmental adaptability

> » Efficiency in water use

> » Early onset of biological nitrogen fixation

> » Strong N accumulation rate

> » Timely release of nutrients

» Insensitivity of photoperiod

» High production of seed

» High viability of seed

» Ease of incorporation

» The ability to cross-inoculate or respond to inoculation

» Pest and disease resistance

» High N sink in sections of the underground parts.

Table 13.14: Criteria for Selection of Green manure

Criteria	Effects
High biomass production	Mobilization of soil nutrients into vegetation; weed suppression;
Deep rooting system	Pumping of moist and/or leached nutrients from soil layers not absorbed by main crop roots
Fast initial growth	Fast soil protection for efficient soil conservation; weed suppression;
More leaf than wood	Simple organic matter decomposition
Low C:N ratio	Enhanced availability of nutrients for successor crops; easy handling during-cutting and/or incorporation into soil;
Nitrogen fixing	Increased availability of nitrogen
Good affinity with mycorrhiza	Mobilization of phosphorus leading to increased supply of crops
Efficient water use	Effective utilization of water

Advantage of green manuring

» Has a beneficial effect on the physical and chemical properties of the soil.

» Helps to preserve the level of organic matter in arable land.

» Serving as a source of food and energy for microbes multiplies exponentially, it only decomposes GM and results in the release of plant nutrients in the appropriate sources for use by crops.

» Improves aeration in rice fields by inducing algae and bacterial surface film activities.

» Additional use as a fruit, feed and fuel supply.

Soil structure and tilth improvement

» Green manure makes up the soil structure and improves tilth.

» Promotes the creation of crumbs in hard soils that contribute to aeration and drainage.

» Increases the water holding capacity of the light soil.

» Shape a canopy covering over the soil, minimise the temperature of the soil and prevent rain and water from erosing.

Fertility improvement of soils

» Absorb the lower layer of nutrients and drop them on the soil as ploughed.

» Prevents the leaching of nutrients to lower levels.

» Harbour N fixing bacteria, rhizobia in root nodules and ambient N fixing (60 to 100 kg N / ha).

» Increase the solubility of lime phosphates, trace elements, etc. by soil microorganism activity and organic acid production during decomposition.

Improvement in crop yield and quality

» Increases grain yields to 15 to 20 per cent.

» Increased vitamin and rice protein content.

Amelioration of soil problems

» *Sesbania aculeata (daincha) added continuously to sodic soils for four to five seasons increases permeability and tends to recover.*

» *Argemone mexicana & Tamarindus indica has a buffering function when added to sodic soils.*

Pest control

» Pongamia and Neem leaves have insect control effects.

Limitations of green manure

» Under rain-fed conditions, it is believed that adequate decomposition of the incorporated green manure will not occur if ample rainfall is not obtained until the green manure crop has been buried.

» Since green manure for the loss of *kharif* wheat, the use of green manure will not always be economical.

» Often the cost of green manure crops can be greater than the cost of industrial fertilisers.

» Sometimes it causes cancer, flies, and nematode problems.

» The green manure crop will fail if there is not enough rainfall.

No-Cost Inputs

No-cost inputs are those inputs that cost nothing or cost the bare minimum but have high benefits. The following are important non-cost inputs that are useful for organic farmers:

1. Indicator plants

2. Use of planting calendar

3. Homa therapy or agnihotra

Indicator Plants

If a nutrient is not available in sufficient quantities in the soil or is not supplied to the plant in sufficient quantity, the signs of deficiency mentioned below may be seen to a greater or lesser degree, depending on the magnitude of the deficiency. However, certain plants have been shown to be exceptionally useful as markers of specific deficiencies. These plants are particularly prone to a particular deficiency and the signs of deficiency such as slow growth and changes in the colour of the leaves are seen more strongly by certain indicator plants.

Table 13.15: A list of such indicator plants suitable to indicate various deficiencies is given below.

S. No.	Deficient element	Indicator plant
1	Nitrogen & Ca	Cauliflower, cabbage
2	Phosphorus	Rapeseed
3	Copper	Wheat, oats
4	Molybdenum	Cauliflower
5	Magnesium	Potato, cauliflower
6	Zinc	Citrus, cereals, linseed
7	Manganese	Oats, sugar, beet, potato
8	Boron	Sugarbeet, cauliflower
9	Potassium	Potato, cauliflower, broad beans
10	Iron	Cauliflower, cabbage, oats, potato

In addition, sunflower and crotons are markers of soil moisture stress. Farmers can irrigate the crop by looking at the wilting signs of these plants.

Use of the planting calendar

The patterns of life of all living creatures are incorporated into the universal flow. The scientific world of science does not acknowledge the effect of these cosmic rhythms and constellations on life forms. Yet human life, as well as animal and plant life, is all heavily dependent on the rhythms of the earth. In the same way, plant and animal life is often affected by the syndic relationships of the sun, the earth, the moon and other planets. On the basis of such conditions, the calendar of plantings shall be planned for agricultural operations at different timings of the year.

The moon opposite to Saturn

Occurs approximately once in 29.5 days.

Activities to be undertaken:

1. Seed sowing, transplanting, grafting, pruning and layering.
2. Spraying BD 501 (cow horn silica) to manage pests.
3. Spraying liquid manures and foliar sprays.

Full moon

Occurs every 29.5 days

Activities to be undertaken:

1. Sow seeds two days before sowing.
2. Apply liquid manures and CPP (cow pat pit) manure.
3. Spraying bio pesticides to control pests and diseases
4. Drenching the animals to remove internal parasites (48 hours before).

New moon

Happens once in 27.5 days

Activities to be undertaken:

1. Avoid sowing seeds.
2. Cutting timber.

Ascending periods

The moon moves in an arc from east to west and when this arc gets higher, the moon is ascending.

Activities to be undertaken:

1. Sowing of seeds.
2. Spray BD 501.
3. Spray liquid manures and CPP.

Descending periods

The moon moves in an arc from east to west and when this arc gets lower, the moon is said to be in descending phase.

Activities to be undertaken:

1. Transplanting of seedlings.
2. Spraying BD 500 (cow horn manure).

3. Making and spreading compost.

4. Pruning trees.

5. Land preparation activities.

Nodes

These are the days when the moon passes the sun's path. It creates negative influences on the growth of plants. Avoid all agricultural activities during nodes.

Apogee

The moon's orbit around the earth is elliptical. The point where the moon is furthest away from the earth is called its apogee.

Activities to be undertaken:

1. Planting potatoes.

2. Irrigating the field.

Perigee

The moon travels in an elliptical path across the earth. The position where the moon is closest to the earth is called the perigee. Spray biopesticides to manage pests and diseases.

Seed and fruit days

These days affect the growth of seeds and fruit crops and are good for sowing and harvesting, e.g. paddy, wheat, brinjal, bhendi and tomato.

Root days

These days affect the growth and development of root crops and are excellent for sowing and harvesting: potatoes, carrots, beet root, etc.

Flower day

These days affect the growth and development of the flowers and are ideal for sowing and harvesting: cut flowers, cauliflowers, roses, jasmines, etc.

Leaf days

These days aid in the growth and development of leafy vegetables and are excellent for sowing and harvesting: green leafy vegetables, cabbage.

Homa Therapy or Agnihotra

Homa is a Sanskrit term used synonymously with yajna or havana. Yajna is the scientific term of the Vedic Bio-Energy Science, which denotes the method of eliminating the toxic conditions of the atmosphere through the fire. This includes curing and purifying the world with fire as a medium. You heal the atmosphere, and the healed atmosphere will heal you. This is the fundamental concept of the Homa Therapy. This knowledge can be used in agriculture, the environment, medicine, psychotherapy, biogenetics, etc.

Agnihotra is the basic homa to all the Homa Fire practises mentioned in ancient Vedic science. It's tuned to the biorhythm of sunrise and sunset. The process involves preparing a small fire in a copper pyramid of fixed size and putting some grains of rice in the fire at sunrise and sunset, accompanied by the chanting of two simple mantras.

Farmers in more than 60 countries are taking Homa Therapy. There are many reports from India and abroad that claim that the use of Homa Therapy in agriculture improves degraded land, controls pests and diseases and improves production quantity and quality.

Chapter - 14

Tissue Culture for
Hi-tech Horticulture

Introduction and Definition

Agriculture today is on the verge of a technological revolution, in a way that has never been seen before. When we reach the new millennium, one of the main innovations that comes to mind is the advent of biotechnology, which provides some of the greatest possibilities and answers to some of the uncontrollable challenges we face.

Biotechnology is a community of inventions that have two aspects in common; they control living cells and their proteins, and they have a broad variety of applications that can change our lives. Major biotechnology techniques include genetic engineering, cell culture, tissue culture, bioprocessing, protein engineering, etc.

Plant Tissue Culture, Micro Propagation or Cell Culture is a method used to grow selected plants of proven suitable agricultural quality in a large number of small plant parts in a fairly short period of time. It is a method of rapid spread under controlled disease-free conditions. An entire crop population of premium quality can be produced from a single elite specimen plant. Based of the species in question, the initial tissue can be extracted from the shoot tip, leaf, lateral bud, stem or root tissue of the mother plant. Ex-plants from selected mother plants are developed and multiplied under 'In-Vitro' conditions, providing an optimum pre-requisite for plant development. Such ex-plants undergo initiation, multiplication and rooting procedures for the conversion of a cell into a full-fledged plant. Such ready-made seedlings are then cured in temperature regulated green houses or polyhouses. Depending on the species, the plants will be ready for field planting. This technique of plant propagation significantly reduces the labor and space requirements for the production of new varieties and may also significantly increase the rate of spread.

Plant propagation and establishment methods are of particular interest to our country and work on a wide range of vegetables, fruit crops and trees is ongoing. Several scientists have been working to expand the usage of tissue culture to make plant species economically valuable. For example, experiments on crops such as coconut, date palm. Cashew, Mango, Orange, etc. are produced in different research laboratories.

With the advent of plant tissue technology, it is now possible to propagate fine varieties of flowers, forests and fruit trees by small plantlets. The commercialisation of these crops has also taken place. Ornamental crops, orchids, carnation, gladiolus, gerbera, anthurium, etc. have been commercially grown. Fruit varieties, bananas, sugarcane, etc. have been commercially produced. In the trees of the forest like Teak, Eucalyptus, etc. have been commercially produced. Medicinal plants are also being tested by a method of tissue culture and will soon be ready for commercial planting.

Biotechnology is an area with enormous potential to solve basic food, fiber, fuel and medicine problems, particularly in developing countries in Asia.

Indian Scenario

In the 1980's, when all these countries developed millions of plants, India had just four commercial tissue culture laboratories. Eventually, the laboratories increased, but were unable to generate the quantity required by the agriculture and horticulture industry.

Most large plant tissue culture laboratories started operations in the 1990s. Plant tissue culture is currently well studied, evaluated and recognized in India. India has accomplished a breakthrough in this methodology by undertaking research and development with well-equipped testing laboratories such as the Indian Council for Agriculture Science, Delhi; Indian Institute of Horticulture Research, Bangalore and National Chemical Laboratory, Pune. Now the need is to make this technique strong in the commercial area of production.

Despite of the support of the national level Research Institutes, Agricultural Universities and Government Agriculture Department, commercial tissue culture is still facing multifarious problems.

Advantages of Plant Tissue Culture

» **Large reproduction of elite clones:** micropropagation enables the development of vast quantities of plants from tiny sections of the mother plant. It requires a fairly limited period to produce plants. Depending on the species under production, a single ex-plant may be multiplied by several thousand plants in less than one year.

» **Elimination of diseases in planting material:** another reason for which plant tissue culture is ideally adapted is the growth, conservation and mass propagation of particular disease-free plants. The concept behind pest-free plant indexing is closely linked to the concept of using tissue culture as a selection system. Plant tissues known to be free of disease under consideration (viral, bacterial or fungal) are physically selected as tissue culture explants. Tissue culture could be a useful way of circumventing or eliminating diseases that may arise in a stock plant.

» **Plant development by tissue culture:** the production of superior varieties of agricultural crops is possible by means of a system of tissue culture, otherwise not achievable by traditional methods of plant breeding.

» **True to type of production:** a large number of plants true to type could be propagated within a short time and space and throughout the year. For eg, two to four lakhs of tissue-grown plants could be propagated from a single bush or rose to 10 to 15 plants through traditional means. It may also take about two to four months for healthy planting material to be produced by means of tissue culture, while a minimum of six to eight months is required for most species by the most recent method of plant propagation.

» **Higher yields:** Tissue culture Plants may have increased branching and flowering, increased vigor and higher yield, mainly due to the possibility of disease elimination.

» **Beneficial where traditional propagation is difficult:** the system can succeed in propagating plants where seed or traditional propagation is not feasible or is difficult or undesirable.

» **Efficient method for saving space and energy:** this method saves the farmer's space and energy. For example, in the traditional process, plants are grown in an open farm involving a region of around 25.000 m, with the same amount of plants needing just 10 m2 if they are grown in a tissue culture laboratory.

> » **Flexible method:** the versatility of nurseries can be enhanced. As capital investment in the mother plant is reduced to almost zero, it may be easier to adapt to changing conditions. In fact, a smoother manufacturing system is feasible thanks to improved plant uniformity and supply of mass at any moment.

> » **Innovation of new species:** Tissue culture may be used for the propagation of new varieties.

Conceptual Foundation of Plant Tissue Culture Technology

Plant tissue culture refers to the production in 'In-Vitro' (Vitro-Glass) of certain aspects of the plant, including a single cell, tissue or organ, under disease-free conditions on a nutrient medium.

In the life cycle of every individual, two sets of opposite sex combine to create a single cell-zygote. This single-celled zygote is the root of the whole multicellular and multi-organic body of the higher organism. In a flowering plant, for example, structures as functionally complex as underground roots, green leaves and flowers all arise from a single-celled zygote across millions of cell divisions.

Theoretically, therefore, all cells in the plant body, whether in flowers, tissues or root tips, should have received the same genetic material as was originally present in the zygote. Any other influences ought to be superimposed on the genetic properties of the cells, which give rise to this large variability represented by genetically similar cells. The cycle of variance is called differentiation. In fact, this differentiation is preceded by certain cellular and subcellular changes. The question that arises at this stage is whether the cellular changes underlying the differentiation of different types of cells are permanent and irreversible or whether there is merely an adaptive change to the functional needs of the organism in general.

During the normal life cycle of a plant, it is believed that events leading to differentiation are of a permanent nature. However, Vochting studies on polarity in cuttings (1878) indicated otherwise. He found that all cells along the duration are capable of developing roots as well as shoots, but their density is dictated by their relative location of the cutting.

However, the best way to answer this question and to understand more about the interrelationship between the different organ cells and the different organs of the organism would be to remove them from the influence of their neighboring cells and tissues and to grow them in isolation on the nutrient media. This led to the creation of a new biology branch called 'Cell and Tissue Culture.' It refers to both plant and animal cells. Plant tissue culture has gained a range of functional uses in agriculture, horticulture and forestry. It is becoming increasingly popular as part of the recent field of biotechnology.

Applications of Plant Tissue Culture

In the early fifties, it was found that plant cells are capable of chemical modulation in the medium through which they can be manipulated to shape ordered structures and full plants. This finding is considered to be very significant for the implementation of methods of cell and tissue culture to solve a variety of problems relevant to agriculture, horticulture and plant breeding.

1. The technique provides a means of rapid multiplication of desirable and rare plants. 20,000 plants / year / bud in turmeric, 1,000,000 plants / year / bud in Eucalyptus were found.

2. As the tests have demonstrated, virus-infected plants often produce several safe stocks as well, which can be collected by splitting shoot tips for their in vitro propagation. It has resulted in succession outcomes for Strawberries and Sugarcane.

3. The production of haploids by anther culture technique has a possible significance in specific and applied genetics and in plant breeding. In the last 20 years, the method has been effectively applied to nearly 20 plant species, including several commercial plants.

4. The embryo culture has proven successful in resolving the dormancy of the seed. It is often used to develop productive cross-breeding plants, which usually fail due to the death of immature embryos. In the case of Jute and Rice, the tests were successful.

5. The cultivation of embryo tissue is often used for the cultivation of unusual plants. In certain studies, coconuts formed fluffy, stable and fatty tissue instead of liquid endosperm. They are uncommon and very costly, eaten only at exclusive banquets in the Philippines. Coconut seeds do not germinate in natural circumstances. Using the technique of in-vitro cultivation of excreted embryos, the development of plantlets has been effective.

6. A more significant application of embryo culture is used in the development of certain unusual hybrids. It is possible to develop full hybrid plants by

embryo culture. This approach has been used profitably for many interspecific crop crosses such as tomato, papaya and cotton.

7. Single plant cells may be separated and cultured. It assists in the development of mutants in relation to crop improvement, such as tobacco, Datura, etc. The method is also effective in the processing of other chemical compounds in the field. In certain instances, cell cultures contained 20 times more chemical content than roots.

8. Recently, tissue culture is used in protoplast culture of various plant varieties and these protoplasts are used for somatic hybridization.

9. A few high-performance crop varieties have been commonly introduced, resulting in the absence of a significant number of older varieties. The forests, which are host to the wild races of much of the crops, are being cut on a large scale. Therefore, tissue culture should be used to maintain germplasm, i.e. tissue conservation of certain plants can be achieved in the same manner.

Important Steps in Plant Tissue Culture

1. **Medium preparation and sterilization in autoclave:** In the industrial facility, the material is processed for initiation, subculture and rooting purposes. The medium consists of nutrients such as Micronutrients, Macronutrients, Vitamins, Irons and growth regulators necessary for plant development. The material is sterilized at a temperature of 15 pounds in the electrically powered autoclave. This sterilization is necessary to avoid bacterial contamination. This fluid is then poured into the culture bottles and these bottles are sterilized again in the autoclave. The substance is then cooled and primed for inoculation.

2. **Selection of the mother plant and sterilization:** the industrial laboratory determines growing plant species are to be multiplied in the facility. Mother plants are then selected from free areas of the virus. Healthy, disease-free plants and virus-free plants are selected as mother plants. The real plant portion, which is called the ex-plant, is selected for inoculation. This portion of the plant is cleaned with water, liquid soap and antiseptic solution. This ex-plant is washed again with a chemical solution to avoid fungal contamination from the field environment. This is carefully cleaned with distilled water to eliminate the traces of the toxins.

3. **Inoculation of explants in a sterilized medium with known composition:** Inside the laminar flow facility, the explant is again handled with a disinfectant-Chemical

named Sodium Hypochloride. Following this procedure, the ex-plant is washed three times with double distilled water to eliminate the chemical residues. The explants are inoculated in a sterilized medium. The inoculation takes place in the inoculation room at the Laminar Air Flow station. This ensures a steady air stream, which leaves this bench clean of the possibility of pollution. The inoculation is performed by professional technicians in the laboratory.

4. **Shifting to the growth room:** Inoculated crops are then transferred to the growth room where the usual temperature of 15-20°C is preserved with the aid of air conditioners and clean air of Class 1000. The artificial lighting system is also built for the development of the plant.

5. **Shifting of cultures for subculture:** After three to six weeks of inoculation, the inoculated explant displays development in multiple shoots. Such shoots are transplanted to the subculture or multiplication system for further development of the plants. Separate growth medium is designed for subculture. Many subculture processes are conducted in the ex-plant industrial processing laboratory.

6. **Separation of in-vitro shoots and rooting:** In the laminar airflow stations, the shoots are separated and these completely grown shoots are moved to the rooting medium for root production. Based on the insect, rooting takes one to three weeks.

7. **Transfer to green house for hardening:** Rooted shoots are collected from the laboratory and sent to the green house for hardening. During the process of hardening, the plants are first placed in the humidity chambers for acclimatization and then moved to the green room.

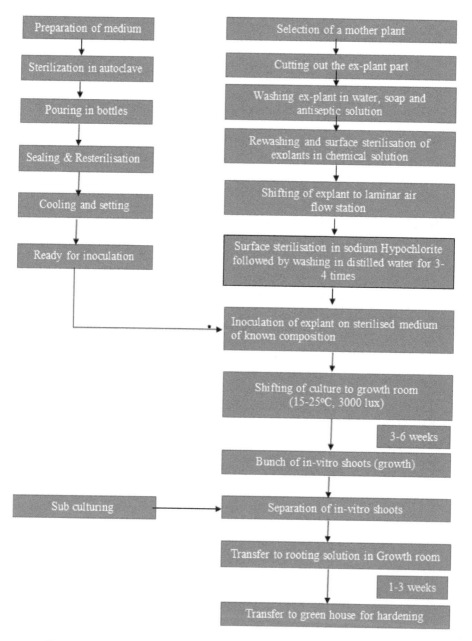

Fig.14.1: Flow chart of plant propagation by tissue culture method

Commercialization of Plant Tissue Culture technique

In Western Europe, industrial micro-propagation started as early as the early 1980s, with a total of two hundred and forty industrial plant tissue culture laboratories generating millions of plants a year by the end of 1988. Also Holland started industrial micro-propagation in the 1980s, and by the end of 1988 Holland developed sixty-two million plants with the aid of seventy-seven tissue culture laboratories. Only Israel had five large tissue culture laboratories growing five million plants a year by the end of 1988.

Several commercial tissue culture laboratories have also been set up in other countries and millions of plants have been developed for their market. Such nations included Poland and the Yugoslavia, and Soviet Union etc. Throughout America, experimental micro-propagation started in 1965, and soon there were only a hundred industrial tissue culture laboratories.

Micropropagation technology parks offer following services to the tissue culture business:

1. **Transfer of technology:** transfer of validated technologies to customers, preparation, deployment and introduction of technology at the customer's location.

2. **Contract research:** production of the newest crop method, refining of the current procedure.

3. **Technical assistance for plant production:** there is an indigenously built, highly sophisticated laboratory and green house for the cultivation of large-scale plants. A large community of scientists and highly qualified personnel are at the service of the unit who offer professional support for the development of planting programs or field trials.

4. **Training of personnel:** training programmes are based around general training for plant tissue culture and advanced training for particular plants of interest.

5. **Establishment and commissioning of a tissue culture laboratory and a greenhouse:** construction of a laboratory and a green house, equipping a laboratory and a green house.

6. **Advisory consultancy for the general operation of the laboratory:** consultancy

given for the administration of the tissue culture facility, which covers all things, from the washing of glassware to the processing and hardening of plants in the green house.

7. **Turn-key project based on the client's need:** Each of the above-mentioned resources may be taken individually or in conjunction when needed. The Turn-key proposal, which involves all of the above listed resources together and all other relevant issues or difficulties, is often presented.

Conclusion

Propagation through tissue culture provides strong market opportunities for ornamental plants, vegetables and fruit plants, where the benefit of the crop is high. In India, the technique of tissue culture has reportedly been effective in more than a hundred species of plants. It has been reported that more than three hundred and fifty million tissue-cultivated plants are produced annually in India using a tissue culture process.

Plant Tissue Culture has come to remain as a resource for plant biology. Plant Tissue Culture has the ability to address experimental biology issues that are often impossible to address by traditional methods. This methodology will play a very influential role in the genetic manipulation, reproduction and afforestation systems in the immediate future. Tissue culture methodology is a plus for the agriculture and horticulture sector because of its multiple advantages. Tissue culture will yield a variety of good, virus-free and true-totype plants. The downside of this approach is that such plants may be cultivated at any point of the year that eliminates the seasonal planting of farmers.

Disease-free plants, compounded by tissue culture, achieve higher yields than contaminated plants. This use of biotechnology can be used with both conventional and modern varieties. Carefully controlled output will insure the delivery of healthy plantlets.

Tissue culture operation is taking the shape of an industry, as many farmers are planting tissue culture plantlets; agro-traders are buying and selling tissue culture plantlets, while others are either exporting plantlets or growing plants grown by tissue culture, particularly flower varieties such as roses, anthuriums and gerberas are enjoying high income. Tissue science is the greatest breakthrough in plant breeding. The agriculture and horticulture industry will make good use of this strategy to achieve the maximum standards at national and international level.

In India, tissue culture is increasingly becoming a popular tool for the cultivation of new and rare organisms and difficult-to-propagate plants. A brand new market is growing from a few small labs a few years back. Currently, the market for micro-propagated plants is greater than the availability of certain products. Few growers concentrate only in the micropropagation of plantlets, leaving the growing-on, i.e. the hardening process to others. Many growers are incorporating a tissue culture laboratory into their overall activities.

Although more plant tissue culture laboratories are developing, several of the laboratories are being closed for different purposes, and several of the current laboratories have been found to be complaining about the challenges they face. Commercial usage of plant tissue culture technology has a tremendous business potential if tissue culture laboratories are free from their constraints, especially financial and marketing constraints. These are the businesses who will make it possible for India to stay self-sufficient in the area of agricultural development. Finance is at the core of every company, financing would also have to be carefully managed and supervised in the tissue culture industry. The prospective entrant will carry out a comprehensive review of the financial feasibility and market potential of the tissue culture undertaking. This is the need of time in the sense of the age of globalization and the rise of the World Trade Organization.

Chapter - 15

Production Technology under Hi-tech Horticulture

Production Technology of Tomato Under Greenhouse

Scientific name	:	*Solanum lycopersicum* L.
Common name	:	Tomato, Tamatar
Origin	:	Western South America and Central America

The development of varieties / hybrids from both the public and private sectors has played a significant role in the enhancement of crop production. Greenhouse tomato development has gained a great deal of publicity in recent years, partially due to a new surge of interest in sustainable crops. The appeal is focused on the belief that greenhouse tomatoes are more productive than traditional horticultural crops.

Soil

Tomatoes can be grown in a wide range of soils as long as the drainage and physical structure of the soil is strong. Optimum soil pH varies from 6.0-6.5, but crops can perform well in soils with a pH of 5.0-7.5. As the pH decreases below 5.5, the supply of magnesium and molybdenum drops below 6.5, and copper, manganese and iron are deficient. Many greenhouses in India use soil-based media for crop growth. The land-based medium will consist of 70 per cent red soil, 20 per cent well-decomposed organic matter and 10 per cent rice husk. The raised beds with a height of 40 cm and a width of 90 cm are made for effective cultivation of the crop. It is now a necessary activity for all greenhouses. The growing medium of the greenhouse is usually pasteurized every year. Such elevated pace is sometimes needed due to the proliferation of disease-causing species in the greenhouse. Formaldehyde is a compound widely used to sterilize the soil media. Formaldehyde (37-41 percent) used for sterilization will be mixed with water in a proportion of 1:10 and soaked at a rate of 7.5 liters per 100 m², i.e. 75 liters of formaldehyde per 1000 m² of polyhouse.

Upon soaking, the soil substrate shall be filled with plastic film or black polyethylene sheet and all drainage areas shall be blocked. Three or four days after formaldehyde treatment, the polyethylene cover is removed and the product is adequately treated or fully absorb the excess / remove accumulated formaldehyde vapours prior to transplantation.

Climate

Temperature is the primary factor that affects all stages of plant development:: vegetative growth, flowering, fruit setting and fruit ripening. Growth includes temperatures of between 10°C and 30°C. Temperature and light intensity influence the germination, vegetative growth, fruit environment, pigmentation and nutritional value of the fruit. The minimum temperature for seed germination ranges from 8 to 10°C. The night temperature is a vital element in fruit settings with an optimal range of 16°C to 22°C. Fruits are not set at 12°C or below. The fruit collection is also greatly decreased when the average maximum day temperature is above 32°C and the average minimum night temperature is above 22°C. Nevertheless, several varieties are now available which can set fruit above the critical temperature limits and are called hot set and cold set cultivars, respectively. Light intensity is one of the key factors influencing the amount of sugar contained in the leaves during photosynthesis, and this, in effect, influences the number of fruits that the plant can sustain and the total yield. Optimum relative humidity in glasshouse crops varies from 60 to 80 per cent. Under hydroponics, 75 per cent to 85 per cent are common night to day relative humidity. Moisture outside this range can inhibit the fruit setting phase.

Choice of varieties

Selection of the most suitable cultivar is a requirement for successful growing of tomatoes in a greenhouse. The growth pattern of the greenhouse cultivars is indeterminate and the plants can exceed a length of 10 feet or more during the growing season. The first step in the production of every crop is to select the best variety. It is wise to continue with the greatest ability rather than restrict yourself by using inferior seed, even though the initial capital is saved.

Thousands of varieties of tomato are available on the market, but only a handful are appropriate for greenhouse growth. These are mainly indeterminate varieties / hybrids, developed exclusively for greenhouse growth.

Criteria for the selection of variety for greenhouse cultivation:

» The desired fruit size.

» Susceptibility to the disease.

» Lack of physiological issues, i.e. cracking, cat face, blossom end rot.

» Yield uniformity.

» Competition from the market.

Tomatoes grown in greenhouses are typically classified into the following categories:

1. Beefsteak cultivars

Beefsteak cultivars grow large slicing fruit of about 180 to 250 g in weight. The fruits are normally picked separately and are usually lined with calyx. These cultivars are very common among greenhouse growers in almost all European countries and the United States.

2. Big fruited cultivars

The fruits of these cultivars typically weigh between 80-150 g and come in small to medium clusters. While several cultivars have been produced for greenhouse cultivation in Europe, the USA, Turkey, Israel, etc., but in India, there is a limitation on cultivars bred exclusively for controlled cultivation, with the exception of a few cultivars such as Pant poly house tomato-1, Pant poly house tomato hybrid-1. Otherwise, indeterminate cultivars from both the public and private sectors are commonly grown under protective conditions, the list of which is given below.

Table 15.1: Indeterminate/ Semi-determinate Hybrids/ Varieties in India

Public Sector	Arka Abhijit, Arka Ananya, Arka Ahuti, Arka Rakshak, Pusa Hybrid-2, Arka Shreshta, Pant poly house tomato hybrid-1, Pusa Ruby, Pusa Uphar, Arka Samrat, Pant poly house tomato-1, Azad T-5, COTH 1, COTH 2, TNAU Tomato Hybrid CO3.

Private Sector	Anup, Avinash-2, Heemsohna, Trishul, To-848, Rakshita, Naveen 2000+, Avtar, ARTH-128, Naveen, Vishwas, NTH-2004, NTH-2005, NTH-2008, Saberano, NS-4266

3. Hand type or cluster type

These cultivars grow fruit in clusters of four to seven or more and are typically harvested and sold with a complete cluster of fruits from breaker to ripe level. The weight of the single fruit of such cultivars is roughly 50-70 g.

4. Cherry tomato

This group is attracting a lot of attention from farmers, as buyers are now choosing these cultivars for table purposes. These are very small in size, round or oval in shape, and the average fruit weight is 12-20 depending on the cultivar. Cultivars such as NS574 (904), NS575 (907), NS 6438 (Namdhari Seeds Pvt Ltd.), Solan Red Round (UHF, Nauni-HP), Laila, Sheeja, Roja, Ruhi [Known you seed (India) Pvt. Ltd.].

5. Coloured tomato

More recently, cultivars with an excellent flavour and abundant in antioxidants and vitamin A have gained popularity among greenhouse growers at a better quality. Brown and yellow coloured cultivars are eligible for cultivation in this region.

Sowing time, nursery raising and spacing

Tomato can be cultivated during the year under greenhouse conditions, but careful attention must be paid to it throughout the off-season. The nursery for greenhouse tomatoes grows under a safe system, usually in soils with fewer media in plug trays to produce disease-free and mostly virus-free seedlings. Ingredients of soil less media viz., cocopeat, vermiculite and perlite combined in a ratio of 3:1:1 before filling pro trays or plug trays. In general, plug trays with 98/104 plugs or small plugs are recommended for the sowing of tomato seeds.

Fertilizer management

Fertility is potentially one of the most frustrating topics for growers of greenhouse tomatoes. The keys to a good nutrition program include:

» Using fertilizers specially designed for greenhouse tomatoes.

» Know how much of each item of fertilizer is required.

» Know how much is being used.

» Test the rate of electrical conductivity (EC) and pH.

» Observe signals that plants may be deficient or have an abundance of a nutrient.

» If necessary, track the state of plant nutrients by taking samples annually for tissue examination.

The crop is fertilized with N:P:K at a rate of 25:12.5:12.5 kg per 1000 m² along with the manure of the FYM (2t/1000 m²). Fertigation was planned after 10-15 days of planting at weekly intervals. The typical micronutrient dose (G-5) at a rate of 5 kg per 1000 m² was used before the crop was transplanted.

Table 15.2: Fertilizer management

Crop Duration	Application ratio of fertilizers		
	N	P	K
1st Growth Period (Up to 30 days)	2	3	1
2nd Growth Period (30-60 days)	1	2	2
3rd Growth Period (60-90 days)	1	1	3
4th Growth Period (90-120 days)	1	1	2
5th Growth Period (120-150 days)	1	1	1
6th Growth Period (150-180 days)	1	1	1
7th Growth Period (180-210 days)	1	1	1

Irrigation management

If the drippers are at a distance of 30 cm with water discharge rate of 2 lph, implement the following irrigation plan for improved results.

Table 15.3: Irrigation schedule

Crop Stage	Time of operation of drip system / irrigation (minutes)	Frequency of Irrigation		
		Summer	Winter	Rainy
Upto Fruit Setting	60	Alternate Day	Every 4th Day	Every 4th Day
fruit Setting to First Harvesting	75	Alternate Day	Every 4th Day	Every 4th Day
first Harvest to one week prior to last harvest	60	Alternate Day	Every 4th Day	Every 4th Day

Training and pruning

For improved development, prune tomato plants to a single stem, two stems or three stems by cutting all lateral shoots, commonly referred to as 'suckers.' Suckers are developed at the stage where each leaf comes from the main stem just above the leaf petiole (stem). Allowing all suckers to grow and bear fruit would increase the overall number of fruits, but they would be limited and of poor quality. It is best to have one main stem(s) that produces fruit because it can yield bigger, more uniform and higher quality fruit.

Removing suckers once a week should keep them under control. It is best to leave one or two of the smaller suckers on top of the vine. Then, if the plant gets injured and the terminal breaks down, one of these suckers will be allowed to expand and become a new terminal. Generally, eliminate any sucker that is longer than 1(one) inch.

Plants are trained vertically around the support wire to maximize the maximum capacity of indeterminate varieties. This will be leaned and lowered as the plant reaches the wire height. Lower both plants to the same height so that they don't shadow each other. Repeat this procedure any time the plants grow higher than the cable.

Prune the plant to one of the central stems, wrap it around the support string. Also wrap in the same direction, if you start in the clockwise direction, proceed in the clockwise direction; otherwise, if the plant gets heavy with berries, it will fall down the string and break. Many farmers choose to use plastic clips to attach the plant to a line, either in conjunction with wrapping or to replace wrapping.

Cluster pruning can also increase size and accuracy. This includes extracting small fruit from certain clusters, leaving three, four or five of the best fruit. Next, remove the malformed or deformed fruit. Then, delete the smallest fruit, which is typically the last fruit shaped on each cluster. Pruning is performed by farmers to increase the consistency and yield of tomatoes.

Pollination

Tomato flowers have both male and female portions of each flower. Botanically, these are called perfect flowers. In the green house, the wind is not strong enough to disturb the flowers enough to move the pollen. Also if the greenhouse is ventilated with fans, the air becomes fairly motionless on colder days where the fans do not run.

The ideal temperature for pollination is between 21 to 27 ° C. The maximum relative humidity is 70%. At 80% relative humidity, pollen grains remain together and are not easily dispersed. For a relative humidity of less than 60 per cent over long periods of time, the stigma may be dried so that pollen grains do not adhere to it. Greenhouse tomato farmers can use an electronic pollinator to insure that the fruit is harvested efficiently.

Vibrate each cluster (not each blossom) for about half a second. This procedure can also be conducted manually by softly shaking plants in the morning hours when stigma receptivity and dehiscence are at their height. Nevertheless, specific growth hormones can promote the setting of fruit under conditions otherwise not conducive to the proper release of pollen grains. Various growth hormones along with their role in the development of tomatoes are described below.

Table 15.4: Use of plant growth regulators in tomato cultivation

Common name	Chemicals	Dose (mg/l)	Effectiveness
2,4-D	2,4-Dichlorophenoxy acetic acid	2-5 as seed treatment or whole plant spray	Increase fruit set, earliness and parthenocarpy
4-CPAA	4-Cholorophenoxy Acetic Acid	30	Increase fruit set, parthenocarpy
Cycocel	2-Chlorethyl	500-1000	Flower bud stimulation, pigment formation and increase in fruit set
Ethephon	2-Chloroethyl-phosponic acid	200-500 whole plant spray	Flowering induction, better rooting and plant setting

IAA	3-Indole acetic acid	Foliage spray	For good fruit size and yield
IBA	3- Indole Butyric acid	50-100	Increase fruit set
GA	6-4, Indhydroxy methyl 8 methyl gibbereline	50-100 as foliar spray	Elongate shoot growth
PCPA	Parachlorophenoxy acetic acid	50 As foliar spray	Fruit set under adverse climatic condition

Pest and Disease Management

Insect-pests

» **Serpentine leaf miner** (*Liromyza trifolii* Burgess)

Management

1. Sometimes the occurrence begins at the nursery itself. Therefore, eliminate contaminated leaves at the time of planting or within a week of transplantation.

2. Application of neem cake to polyhouse beds of up to 250 kg / ha during planting and apply again after 25 days.

3. Spray neem powder extract 4 % or neem soap 1 % at 15-20 DAPS.

4. If the occurrence is high, cut contaminated leaves and spray Triazophos 40 EC (1ml) mixed with 7.5 g of neem / l.

5. Prevent regular application of chemical pesticides in protected conditions. At most, one spray of Deltamethrin 2.8 EC @ 1ml / l or Cypermethrin 25 EC @ 0.5 ml / l or Triazophos 40 EC @ 2ml / l may be provided, if appropriate.

» **Greenhouse white fly** (*Trialeurodes vaporariorum*)

Management

1. Use virus resistant hybrids, if available.

2. Raise nurseries under nylon nets or polyhouses in seedling trays.

3. Spray Imidacloprid 200 SL (0.3 ml / l) or Thiomethoxam 25 WP (0.3 g / l) in the nursery at 15 days after sowing.

4. Remove infested plants from the leaf curl as soon as the signs of the disease are expressed. It aims to increased the cause of illness inoculums.

5. Pre-transplant the base of the seedlings with Imidacloprid 200 SL (0.03ml / l) or Thiomethoxam 25 WP (0.3 g / l). If protrays are used to raise the nursery, drench the protrays with the chemicals one day before transplantation.

6. After transplantation, give the requisite sprays of Imidacloprid 20 SL (0.3 ml / l) or Thiomethoxam 25 WP (0.3 g / l) 15 days after planting and do not repeat after fruiting, as this may leave hazardous residues in the crop.

7. Place yellow sticky traps covered with adhesive or sticky glue at flower canopy stage to track adult whiteflies.

8. If the traps show whitefly operation, spray Dimethoate 30EC @2ml / l or neem kernel extract 4 percent (NSKE) or putmia or neem oil (8-10ml / l) or neem soap (10g / l).

9. Rogue the infected plants out with the infection as soon as the signs are detected.

» **Fruit Borer** (*Helicoverpa armigera* Hubner)

Management

1. Spray HaNPV at 250 LE / ha + 1% jaggery with sticker (0.5 ml / litre) at evening when the larvae are young.

2. For adult larvae spay Indoxacarb 14.5 SC @ 0.5 ml / l or Thiodicarb 75 WP @ 1g /.

» **Tobacco Caterpillar** *(Spodoptera lituara)*

Management

1. Collection and removal of egg and gregarious larvae.

2. Spray Spodoptera NPV 250 LE / ha + 1% jaggery with sticker (0.5 ml / litre) in the evening.

3. Use poison baiting: Place 10 kg of rice bran or wheat bran with 2 kg of jaggery by adding a little water in the morning. In the evening, apply 250 g of the formulation of Methomyl or Thiodicarb and scatter over the bed. Caterpillars are attracted to ferment jaggery, feed and kill.

» **Red Spider Mites** *(Tetranychus urticae)*

Management

1. Remove the affected leaves and destroy them.

2. Spray 1 per cent of neem oil / neem soap / pongiamia soap.

3. Spray acaricides such as Abamectin 1.9 EC @ 0.5 ml / l or Spiromesifen @ 1 ml / l or Fenazaquin 10 EC @ 1 ml / l in combination with plant items such as pongmia oil or neem oil (8-10 ml / l) or neem soap (10 g / l).

4. When the incidence is serious, remove and destroy all severely infected leaves followed by an acaricide mixture spray with the botanicals mentioned above.

» **Root-knot Nematodes** *(Meloidogyne incognita, M. javanica)*

Management

1. If available, Use of nematode resistant variety/hybrid.

2. Follow crop rotation with marigold, wherever and whenever possible.

3. Seed treatment with bio-pesticides- *Pseudomonas fluorescens* @ 10g/kg seed.

4. Treatment of nursery bed with T. *harzianum* @ 50 g/ sq. m.

5. 2 tons of FYM enriched with T. *harzianum* and *Paecilomyces lilacinus* use for the management of nematodes in the main field apply per acre before planting, along with 100-200 kg of neem or pongamia cake.

6. Apply Carbofuran 3G @ 1 kg ai/ha at transplanting.

Table 15.5: IPM package for insect-pests in tomato under polyhouse

Raise seedlings in pro trays in polyhouse	
15 DAS (Days after seed sowing)	Spray the plants with Imidacloprid or Thiomethoam.
One day before transplanting	Drench the base of seedlings with Imidacloprid or Thiomethoxam
At transplanting	Apply neem cake 250 kg/ha.

15 DAP (days after planting)	Spray the seedlings with Imidacloprid or Thiomethoxam
25 DAP (days after planting)	Apply neem cake 250 kg/ha
Post flowering and fruiting stage	Monitor for pest like fruit borer, tobacco caterpillar, leaf miner, whitefly and red spider mite.
	Erect yellow sticky traps to monitor whitefly.
	Spray NPV according to the pest.
	Remove leaves severely infected with leaf miner/ red spider mite
	Spray neem seed powder/neem soap for leaf miner.
	Spray synthetic acaricide/botanical in rotation to control red spider mite, spray systemic insecticide/ botanical to control whitefly.

Disease

» **Damping off** (*Pythium, Rhizoctonia, Phytophthora,* etc.)

Management

1. Remove the affected seedlings from trays as soon as the symptoms are visible.

2. Hot water treatment of seeds (at 52oC for 20 min)

3. Treat the seeds with Thiram/Captan /Metalaxyl-Mancozeb @ 2.5-3.0 g/ kg of seeds.

4. Drenching the plug trays 7 days before sowing with Thiram/Captan or any copper fungicide @ 3g / litre of water.

» **Early blight** (*Alternaria solani,* A. *alternata f.sp. lycopersici*)

Management

1. Crop rotation with non-solanaceous host is important for successful inoculum reduction. Healthy seeds can only be obtained from disease-free berries.

2. Field sanitation by plucking the lower leaves and burning the infected crop debris. Minimize the relative humidity of the plant canopy to avoid infection.

3. Spray a crop with 0.3 % mancozeb or 0.1 % carbendazim or 0.2 % chlorothalonil at 10-15 days interval beginning 45 days after transplantation as a prophylactic measure.

4. Two chlorothalonil sprays at 0.2 % or propiconazole (0.1 %) at 8-day intervals are effective against disease, but spraying must be begun soon after the floral portion infection.

» **Late blight** (*Phytophthora infestans* (Mont) de Bary)

Management

1. Do use safe and approved seeds collected from disease-free zones.

2. Infected crop debris and fruit must be gathered and burnt.

3. Preventive sprays of Mancozeb @ 0.25 % have strong control in cloudy, cold and drizzling conditions, but the spray interval should be between 5 and 7 days.

4. One Metalaxyl+ Mancozeb spray @ 0.2 % is very effective when sprayed within two days of infection, but repeated sprays should not be provided.

5. If late blight is present in the pockets, continue spraying with metalaxyl-mancozeb (0.2%)/cymoxanil-mancozeb (0.25%)/dimethomorph (0.1%) + mancozeb (0.2%) immediately and alternate spray schedule with protective and systemic fungicides at 7-8 days of interval based on weather condition

6. Staking of plant reduces all *Phytophthora* diseases of tomato.

» **Powdery mildew** (*Leveillula taurica*)

Management

7. Spraying with karathane (0.1%) or wettable sulphur (3 g/ litre of water) twice at an interval of 10 days helps to control the disease.

» **Bacterial Spot** (*Xanthomonas campestrispv. Vesicatoria*)

Management

1. Summer ploughing to desiccate the bacteria and host.

2. Plug tray nursery raising to avoid seedling infection.

3. Rotation of nursery seedbed and main field. Seed should be collected from disease free plants.

4. Seed dipping in streptocycline solution @ 100 ppm.

5. One spray of Streptocycline @ 150-200 ppm followed by one spray of Copper Oxychloride @ 0.2% in afternoon.

6. One spray of copper oxychloride @ 0.3% after fifteen days of antibiotic application.

» **Bacterial wilt** (*Psendomonas solanacearum* E.E. smith)

Control measures

1. Treat the seeds with hot water at 52⁰C for 20 minutes or with 0.01% streptomycin solution for 30 minutes.

2. Proper crop rotation with non-solanaceaous crops for at least 2 to 3 years reduces the infestation.

3. Apply bleaching powder @ 15-20 kg/ha and 4.5q lime to the land and mix it property to the soil atleast 3-4 week before planting.

4. Proper drainage should be maintained.

» **Leaf Curl Complex** (Virus-transmitted by white fly as well as by mechanical injury)

Management

1. Removal of weed host from field near surrounding areas.

2. Cover the nursery area or plug trays with 40-60 mesh fine nylon net and spray the seedlings with imidacloprid (3.5 ml/ 10lt).

3. Root dipping in Imidacloprid solution @ 4-5 ml per litre of water for one hour during transplanting of the seedlings.

4. Seed treatment with hot water at 50° C or 10% trisodium phosphate solution for 25 minutes.

5. Use barrier crop of taller non-host crops like maize, bajra and sorghum.

6. Apply carbofuran 3G @ 25-30 kg/ha 10 days after transplanting followed by 2-3 foliar sprays with 0.05% dimethoate (Rogar1.5 ml/l) or imidachlorpid (3ml/10 lit) or acetameprid (0.3-0.49/lit) or thiomethoxam (3.5 ml /10 lit) at 10 days interval and alternate the spray schedule with NSKE (0.5%).

7. Periodical sprays of systemic insecticides up to flower setting.

8. Avoid mechanical injury during intercultural operations.

9. Use tolerant varieties.

10. Roughing of infected plants soon after infection at initial stage of growth.

11. Stop cultural practices such as pruning,, training etc. with the hands using tobacco and its products; otherwise go for disinfection of hands either with whey or milk powder solution.

Yield and storage management

The picking of tomato fruits is a continuous process throughout the growing season. Generally, most varieties are able for first harvest in 75-85 days after transplantation. Fruits will be picked preferably early in the morning or late in the evening to prevent post-harvest losses and then graded, packed according to grades. Cherry tomatoes are often picked with stems attached or often alone with calyx attached to the fruit and packed in containers with a size of 400-500 g. A minimum of 25-30 tonnes of big fruit tomatoes and 10-15 tonnes of cherry tomatoes can be harvested from 1000 m² of greenhouse cultivated area.

Production Technology of Capsicum Under Greenhouse

Scientific name	:	Capsicum annuum
Common name	:	Chillies or *peppers*
Origin	:	Southern North America and Northern South America

Capsicum is commonly known as bell pepper or sweet pepper or shimla mirch. It is a member of the solanaceae family and is considered to be a luxurious vegetable used in pizzas around the globe. It is gaining popularity among farmers due to its fast and high returns. Under open conditions, the efficiency and profitability of the crop is low, which reduces the profit margins of the capsicum growers. Bell pepper has attained high-value crop status in various parts of India and is proud of Indian cuisine for its delicacy and good taste, as well as its rich ascorbic acid quality and other vitamins and minerals. Capsicum (Sweet Pepper) rising in India is a recent concern due to increasing demand for fast food at hotels and recent restaurants.

Soil and Climate

Soil

Sweet pepper grows well in loam or sandy loam soils with sufficient water holding capacity. Nevertheless, it may be grown on a broad variety of soils as long as it is well drained. The soil pH for the effective production of pepper should be between 5.5 and 6.8. Most greenhouses for crop production use soil-based technology. The soil medium is 70 per cent red soil, 20 per cent well-decomposed organic matter and 10 per cent rice husk. The raised beds with a height of 40 cm and a width of 90 cm are designed for the effective cultivation of the crop. This is also a necessary activity for all greenhouses. Typically, the root product of the greenhouse is pasteurized annually. However, occasionally it's done for any seed. A change in frequency is sometimes needed due to the proliferation of disease-causal species in the greenhouse. Formaldehyde is a solvent widely used to sterilize the root substrate. Formaldehyde (37-41 per cent) used for sterilization can be combined with water at a rate of 1:10. For drenching Formalin is used at a rate of 7.5 lit per 100 m^2 i.e. for 500-sq.mt polyhouse, 37.5 lit Formalin is required.

After drenching, the soil or root substrate shall be coated with plastic film or black polyethylene sheet. Close all of the air rooms. Replace the polyethylene cover three to four days after formaldehyde treatment.

Climate

Capsicum is a cool season crop and day temperatures below 30°C are optimal for growth and yield. Higher temperature helps in accelerated growth of the plant and impacts the fruit collection. Lower night temperature (20 °C) favors flowering and fruit collection. Shading and fogging are needed during the summer to prevent temperature build-up in the greenhouses. Moderately high RH (50-60 percent) is favored, and can be preserved by ventilation control.

Choice of varieties

The selection of greenhouse sweet pepper cultivars depends on colour, disease tolerance, efficiency and yield. A number of seed companies and retailers sell greenhouse sweet pepper cultivars, and "latest" cultivars are often subject to change as superior cultivars are grown Therefore, it is not recommended that more than one color of pepper be cultivated in the same greenhouse unless it is cultivated under different conditions. The environmental conditions of the different cultivars may be sufficiently distinct to demand that the habitats be handled differently in order to reach optimum yield.

Table 15.6: Selection of varieties

Sr. No.	Colour segment	Varieties
1	Red	Bomby, Inspiration, Pasarella, NS-280, NS-631, NS-632, King Anther, Nun- 3010, Hira, Laxmi, Mamtha, Inspiration, Pasarella
2	Green	Indra, Lario, NS-631, NS-632, NS-274, NS-292, NS-626, Bharat, Mahabharat, Indam K1, Indam K2, Green Queen, Master, Indam Mumtaz
3	Yellow	Orobelle, Bachata, NS-281, NS-626, Yellow Queen, Nun-3020, Tanvi, Golden Summer, Super Gold, Swarna, Bachata,
4	White	White-1
5	Orange	Sympathy
6	Chocolate	Chocolate Wonder

Sowing time and method

Sweet pepper can be cultivated efficiently in greenhouses throughout the year through good management practices. Sweet pepper nursery is grown on soil with less media viz., cocopeat, vermiculite and perlite in pro trays with 98 plugs. Seedlings are planted at a spread of 45 x 30 cm. Nevertheless, when transplanting capsicum seedlings, caution must be taken to prevent close exposure of the collar area of the seedling to the soil.

Cultural and Nursery Practices

» **Nursery raising**

• Good quality seeds are needed to produce better seedlings. Seedlings are grown in pro trays with 98 cells or cavities. Approximately 16,000 to 20,000 seedlings are needed to plant an acre for which 160-200 gm of seed is necessary.

- The pro-trays are filled with sterilized cocopeat and the seeds are sown, with one seed per cell at a depth of 1/2 cm and covered by the same medium.

- The packed trays shall be placed one over the other and lined with plastic sheets before the seeds are germinated.

- A week after sowing, the seeds germinate. The trays are transferred to the net house / polyhouse and are gently watered. After 15 days of sowing, mono-ammonium phosphate (12:61:0) (3g / L) and 22 days after sowing 19:19:19 (3g / L) solution must be drained.

- Spray imidacloprid at 0.2 ml / L and chlorothelonil @1gm / L before planting the seedlings. Start applying approximately 0.3 ml / L of wetting agent per liter of water with each pesticide spray.

» **Land preparation**

The field had to be ploughed extensively, and the soil had to be taken to a good tilth. Fully decomposed organic manure at a rate of 20-25 kg per square meter. It's mixed in with the soil. One application is enough for three capsicum crops to be grown in sequence. The raised beds are built after the soil is taken to a fine tilth. The size of the bed is 90-100 cms long and 15-22 cms wide. Sleeps between 45 cm and 50 cm must be given between the beds.

» **Fumigation**

Crop beds are drenched with 4 per cent formaldehyde (@4 L / m^2 of bed) and covered with a black polyethylene mulch layer. Care should be taken to wear masks, gloves and apron when using formalin. After 4 days of formalin treatment, the polyethylene cover is removed; beds are raked regularly every day to completely remove the accumulated formalin gases prior to transplantation. Formalin care may be replicated after three crop cycles or if needed. Formaldehyde fumigation aims to reduce soil-borne pathogens. Basamid can also be used for soil sterilization purposes.

» **Fertilizer management**

Fertigation schedule for capsicum cultivation under greenhouse is given as under:

25:25:25 kg NPK/ 1000 m.^{2}

Table 15.7: Fertigation schedule

Crop Duration	Distribution ratio of nutrients			Time of fertigation Initiation	Fertigation frequency
	N	P	K		
1st Growth Period (Up to 30 days)	2	3	1	10-15 days of planting	Once a week
2nd Growth Period (31-60 days)	1	2	2		
3rd Growth Period (61-90 days)	1	1	3		
4th Growth Period (91-120 days)	1	1	2		
5th Growth Period (121-150 days)	1	1	1		
6th Growth Period (15-180 days)	1	1	1		

It is also advised to add 0.5 kg of *Trichoderma viride*, Phosphorous Solubilizing Bacteria (*Bacillus megaterium*), *Azotobactor*, *Pseudomonas fluorescens*, 0.4 t vermicompost and 5.0 kg of micronutrients (Grade-5) per 1000 m^2 at the time of planting.

» **Application of neem cake and Microbial Bio-control Agents**

Fifteen days before transplantation, the neem cake must be enriched with bio-agents such as *Trichoderma harzianam* and *Pseudomonas lilacinous*. Neem cake of about 200 kg is powdered and slightly moistened. *Trichoderma harzianam, Pseudomonas lilacinus* and *Paecilomyces chilmdosporia* each of two kilograms are thoroughly mixed with the neem cake. Mixture filled with moist gunny bags or dried grass and left for 8 to 10 days. Evitate direct exposure to sunlight and rain. After 10 days, this enriched mixture of neem cake and bio-agent, along with 600 kg of neem cake, must be added uniformly to one acre field. It is really useful in reducing the issue of soil pathogens and nematodes. *Azospirillum* or *Azoctobacter* or VAM, which is a nitrogen fixing bacteria, may also be applied to the bed.

» **Laying of drip line**

Place one 16 mm inline drip lateral at the centre of the bed with emitting points at every 30 cm interval with a discharge rate of 2 ltr / hr or 4 ltr / hr. Run the drip system to test for consistent discharge at each emission stage before covering the polythene mulch beds.

» **Mulching and Spacing**

Black polyethylene non-recycled mulch film, 30-100 micron thick, 1,2 m wide, is used to cover the planting beds. Holes with a diameter of 5 cm are made on a polyethylene film according to the required width (45 cm x 30 cm). The planting beds are lined with film by pressing the edges of the sheet tightly into the soil. Mulching techniques save energy, suppress weeds and minimize pest and disease infestation which result in higher yields which high-quality growth.

» **Transplanting**

Before transplantation, the planting beds are watered to the field capacity. Seedlings aged 30-35 days are used for transplantation. Care should be taken to insure that no harm has occurred to the roots, when the seedlings are separated from the individual cells of the plant. Seedlings are transplanted into holes created in a 5 cm deep polyethylene mulch film. Upon transplantation, the seedlings are drenched with 3 g / L of copper oxychloride or 3 g / L of captan or 2 g / L of copper hydroxide solution to the seedling base at a rate of 25-30 ml per plant. Watering 10 mulched beds everyday throughout the afternoon, utilizing a hose pipe for a week continuously, is necessary to prevent mortality due to heat trapped by a mulch sheet.

Training & pruning and spacing

The growing point at the top of the plant is removed after 20-25 days of plating, which is known as topping. This method is used for the development of further divisions. Pepper plants initially grow a single stem and, after 7-11 leaves, a terminal flower forms where the main stem divided into two or even three or four. The flower in the first branch is called the crown bud. The bud is not permitted to turn into a fruit and is removed after introduction. Capsicum plants may be raised in a number of schemes, including two stem systems, three stem systems and four stem systems. For two stem systems, after pruning or pinching the other branches, two primary stems are retained on each herb, leaving two leaves or at least one leaf and one flower at each internode. Plants can also be exercised in three or four stems by retaining three or four stems. Within each node, the tip divides into two, giving rise to one strong branch and one weak branch, and the weaker branch is removed. These stems are placed on strings to the main overhead wire at a height of 8-10 feet. The stems are either firmly trellized or wrapped around the strings. Stems can be attached with strings using rings or plastic clips. The crop will grow up to 8-9 feet in height over a period of 9-10 months and after that the plants are topped to prevent stem breakage and to increase fruit production.

Fruit thinning

When there are so many fruits on the plant, it is appropriate to extract certain fruits in order to encourage the growth of the remaining fruits. Fruit thinning is achieved until the fruit reaches the size of the pea. This method is generally practiced in order to maximize the size of the fruit by increasing the efficiency of the harvest.

Irrigation management

If the drippers are at a distance of 30 cm with water discharge rate of 2 lph, implement the following irrigation plan for improved results.

Table 15.8: Irrigation schedule

Crop Stage	Operational Time of MIS (minutes)	Frequency of Irrigation		
		Summer	Winter	Rainy
Upto Fruit Setting	60-75	Alternate Day	Every 4th Day	Every 4th Day
Fruit Setting to First Harvesting	75-90	Alternate Day	Every 4th Day	Every 4th Day
First Harvest to one week prior to last harvest	60-75	Alternate Day	Every 4th Day	Every 4th Day

Pest and Disease Management

Pest Management

» **White or Yellow Mites *(Polyphagotarsonemus latus)***

Management

   » Apply wettable sulphur 80 WP @ 3g / l or any acaricide (directing the spray to the ventral surface of the leaves).

   » Spray pongamia oil (2ml / l) combined with acaricides. Spray neem kernel powder extract 4 percent at a 10-day interval when the pest occurrence is low.

   » Spray acaricides such as Abamectin 1.9 EC @ 0.5ml / l or Oberon @ 1ml / l or Fenazaquin 10 EC@ 1ml / l in combination with plant products such as pongamia oil or neem oil (8-10ml / l) or neem soap (10g / l) as leaves start curling with all precautions.

» **Aphids** *(Aphis gossypi* and *Myzus persicae)*

Management

» Spray Acephate 75 SP @ 1g / l or Dimethoate 30 EC @ 2 ml / l in rotation as required.

» Kill all virus-affected plants and eradicate them.

» **Borers** (Tobacco caterpillar, *Spodoptera litura* and *H. armigera*)

Management

» Spray the NPV particular to the borer type. Inundating activation of *Trichogramma sp.*

» Use marigold as a trap crop (one row of marigold for every 18 rows of chili) to control *H. Armigera*. Collect and destroy the masses of eggs and the immature larvae of *S. Litura*.

» To the *S. Exigua,* Indoxacarb spray 14.5 SC @ 0.75 ml / l or Spinosad @ 0.75 ml / l or Thiodicarb 75 WP@ 0.75 g / l. In certain instances, tomato fruit borer and tobacco caterpillar can even target capsicum under a polyhouse. Follow the tomato management practices.

» Using poison baiting (10 kg of rice flour + 1 kg of jaggery + 250 g of Methomyl 40 SP) for *S. Litura* and repeat baits 2-3 times, if necessary.

» **Root-knot Nematodes** *(Meloidogyne incognita)*

Management

» If available, using nematode resistant variety / hybrid.

» As far as possible, adopt crop rotation with marigold.

» Bio-pesticide crop treatment-Pseudomonas fluorescens @ 10g / kg of seed.

» Care of nursery beds with *T. Harzianum* @ 50 g / sq. m.

» 2 T of FYM supplemented with *T. Harzianum* and *Paecilomyces lilacinus* per acre before planting, along with 100-200 kg of neem or pongamia cake are used for the treatment of nematodes in the main field.

» Apply Carbofuran 3 G @ 1 kg ai / ha to transplanting material.

Disease Management

» **Dieback and Anthracnose** *(Choanephora capsici, Colletotrichum capsici)*

Management

» Disease free seeds should be collected from healthy fruits.

» Screening of diseased fruits must be done after drying of the fruits.

» Seeds should be treated with Carbendazim @ 0.25% during sowing.

» Seedling should be sprayed by Carbendazim @ 0.1% before transplanting. Cut the rotting twigs along with healthy part and burn it.

» Foliar spray of Copper Oxychloride @ 0.3% followed by Carbendazim @ 0.1% at flowering stage. Avoid apical injury during transplanting and also at flowering stage.

» Collect all the green fruits of first setting and consume it. Do not keep these fruits for seed purpose.

» Bacterial Leaf Spot

Management

» One spray of Streptocycline @ 150 ppm alternated with Kasugamycin @ 0.2%.

» Seed dipping in Streptocycline solution @ 100 ppm for 30 minutes.

» **White Rot** *(Sclerotinia sclerotiorum)* **Symptoms**

Management

» Cut the infected plant parts along with some healthy portion in morning and carefully collect in polythene to avoid falling of sclerotia in the field. Burn all these materials away from field.

» Foliar spray of Carbendazim @ 0.1% at flowering stage followed by Mancozeb @ 0.25%.

» **Leaf Blight** *(Alternaria alternata and Cercospora capsici)*

Management

» One foliar spray of Chlorothalonil @ 0.2% alternated by thiophenate-methyl@ 0.1% after 8-10 days.

» Selection of disease free and certified seeds to check the primary infection.

» Foliar spray of Tricel @ 0.2% to maintain crop vigour at 10-12 days interval.

» Field sanitation by burning of infected crops debris followed by summer ploughing.

» Phytophthora Leaf blight/Fruit Rot

Management

» Always use healthy and certified seeds collected from disease-free area.

» Infected plant debris and fruit must be harvested and removed from the field.

» Preventive sprays of Mancozeb @ 0.25 percent have significant influence over cloudy, cold and drizzling conditions.

» One Metalaxyl+ Mancozeb spray @ 0.2 percent is very effective when used within two days of infection, but repeated sprays should not be used.

» Plant elimination reduces the risk of the disease. Rotation, water management and irrigation are cultural practices. Prevent over-cropping and heavy nitrogen.

» **Leaf Curl Complex** *(CMV* and *Gemini Virus)*

Management

» Root dipping of the seedlings in Imidacloprid solution @ 4-5 ml per litre of water for one hour during transplanting.

» Nursery should be grown in nylon net to check the vector infestation.

» Seed treatment with hot water at 50^0 C or 10% tri sodium phosphate solution for 25 minutes. Barrier crop of taller non-host crops like maize, bajra and sorghum.

» Collect healthy seeds from disease-free plants.

» Periodical alternate spray of Oberon @ 1 ml/ 1 with wettable sulpher @ 0.2% and one to two spray of systemic insecticide. Use tolerant varieties

» Initial rouging of infected plants soon after infection and burn it.

» **Sun Scald**

Management

» Adequate fertility and proper water management will help to develop the canopy of leaves and foliage required to protect the fruit from sun scald.

» Timely shading of plants by closing the top shade nets in the greenhouse.

» Poor foliage cover allows the defect to occur. Variety selection may play a key role; small plants may not provide protection as well as more vigorous plants.

Yield

Capsicum harvesting is performed at the green, breaker and colored stage (red / yellow, etc.). It depends on the reason for which it is grown and the gap for the ultimate demand. Throughout India, fruit is picked at a breaker point for long distant markets. It is easier to harvest the colored stage for the local market. The breaker stage is the one where 10% of the fruit surface is colored and when more than 90% of the fruit surface is colored, it is known to be a colored stage.

The harvesting of capsicum fruit stars 60-70 days after transplantation and proceeds up to 170-180 days at a period interval of 10 days. Mature green or 10-15% capsicum-turned fruits weighing between 150 and 250 g are harvested. Keep the fruit harvested fresh and prevent overt exposure to sunlight. All damaged, malformed and bruised capsicums should be discarded. Anyone with dust adhering to the surface should be washed by rubbing a damp soft cloth on the surface. Capsicums can be categorized according to market criteria of the same scale and color quantities.

Table 15.9: The capsicum fruits can be categorized into following grades.

S. No.	Grade	Fruit weight
1	A+	> 200 g
2	A	150-200 g
3	B	100-150 g
4	C	< 100 g

The yield of 10 to 12 t/100 m^2 (10 to 12 kg/m^2, 2.25 to 2.70 kg/plant) can be expected from a crop 8-9-month crop.

Special Tips to Achieve Higher and Quality Yield

» Organic manure / compost added to soil will be enriched with Microbial Bio-control agents such as Pseudomonas fluorescens, Paecilomyces lillacinus, Trichoderma harzianum, etc. and biofertilizers such as PSB, Azospirillum, Azotobacter, etc. to increase soil safety.

» Any harm to the net or polysheet in the framework should be instantly resolved to avoid the entry of pests and diseases.

» The polyhouse / nethouse should have a double door system, which is the safest way to prevent pests and diseases from entering. The doors would ideally be installed away from the roadside

» Seedlings grown in the prostrate must be transplanted within 30-35 days of sowing in the key grown planting beds (1/2 ft above the ground level).

» Regular pruning must be practiced in order to maintain two-thirds of safe branches and to preserve one fruit of reasonable shape and size in each branch and to extract deformed fruits, if any, at a very early point.

» Branches are tightly bound with plastic twine and other ends fixed to the supportive GI board to offer solid support and avoid breaking of branches / fruits.

» Drip irrigation and fertigation schedules should be followed regularly.

» Prevent pests and disease occurrence by taking prophylactic action. Using the correct and recommended quantity and dose of pesticides and insecticides to monitor the occurrence of pesticide using.

» Maintain safety in the green house and dispose of dead, falling and contaminated plant debris / fruits and periodically, ideally every day during the evening hours, after all day operations have been completed. Using reusable plastic bags to store and transport these items to the dump site to avoid the transmission of infection.

» Caution should be taken not to pinch the apical bud and to secure it from a mite infestation.

» Botanicals, Microbial Biocontrol Agents, Biological agents and Biofertilizers can be utilized as an advanced insect, disease and resource management method.

» The yield should be 85-90 percent of 'A' grade fruit (3-4 lobes, 150-180 gms). Deformed and irregularly shaped fruits are pinched at a younger level and fruit with a 50-70 per cent color should be picked, graded and packaged appropriately.

Production Technology of Cucumber Under Greenhouse

Scientific name	:	*Cucumis sativus* L.
Common name	:	Khira, Kakdi
Origin	:	South Asia

Cucumber (*Cucumis sativus* L.) is an edible cucurbit famous around the world for its crisp texture and taste. Cucumber is a particularly versatile crop due to a wide variety of uses, from salads and pickles to nutritional and beauty items.

Soil and Climate

Soil

Cucumbers require lightly textured soils that are high in organic matter, well drained and pH 6-6.8. Adapted to a broad variety of soils, especially early in sandy soils. Cucumbers are fairly resistant of acidic conditions (down of pH 5.5). Greenhouse cucumbers typically develop very well in a wide range of soil pH (5.5-7.5), although a pH of 6.0-6.5 for mineral soils and a pH of 5.0-5.5 for organic soils is commonly agreed as ideal. Many greenhouses utilize soil-based technologies for crop growth. The land-based medium consists of 70 per cent red soil, 20 per cent decomposed organic matter and 10 per cent rice husk. The raised beds with a height of 40 cm and a width of 90 cm are made for effective cultivation of the crop. It is now a necessary activity for all greenhouses. Generally, the root medium of the greenhouse is pasteurized annually. However, occasionally it's done with every plant. This change in frequency is sometimes needed due to the spread of disease-causal organisms in the greenhouse. Formaldehyde is a solvent widely used to sterilize the root media. Drenching root medium with formaldehyde (37-40 percent) combined with water @ 25 ml per liter is the standard practice. For sterilization, formalin should be mixed with water in a 1:10 ratio. Formalin is used to drain at a rate of 7.5 lit per 100 sq. Mt. i.e. 37.5 lit of formalin is required for a 500-sq.mt polyhouse.

The soil or root medium should be covered with black polyethylene sheet or plastic film after drenching. Shut all the ventilation spaces. Replace the polyethylene cover three or four days after treatment with formaldehyde. Two days after the polyethylene sheet has been covered, rake the bed periodically to completely remove the trap of formaldehyde dust prior to transplantation.

Climate

Air temperature is the key environmental variable that influences vegetative production, flowering, fruit production and fruit quality. The growth rate of the crop relies on the ambient temperature of 24 hours, the higher the average temperature of the soil, the quicker it develops. The higher the variation in the day night air temperature, the larger the plant, and the smaller the leaf size. Although optimal growth occurs at daytime and nighttime temperatures of approximately 28 °C, optimum fruit production is attained at nighttime temperatures of 19-20 °C and at daytime temperatures of 20-22 °C. The optimal temperature for safe development need not be less than 18°C. Prolonged temperatures above 35°C should also be prevented as fruit development and quality are compromised at extremely high temperatures.

Warm weather lowers the temperature of the air, particularly during the night, by up to 2°C to promote vegetative growth when it is delayed by heavy fruit loads. This process saves energy since the predominant high temperatures and ideal light conditions assure an average of 24 hours.

Choice of varieties

Variety selection is one of the most significant decisions made during the production process. The cucumber plant has a number of sex types, such as monoecious, gynomonoecious, gynoecious, andromonoecious, androecious, mainly monoecious types, which are found in most varieties. There are also cucumber varieties that grow fruit without pollination. Such varieties are called parthenocarpic varieties, resulting in fruits that are called 'seedless,' while the plant often contains fluffy white seed coats.

The list of varieties of parthenocarpic cucumber recorded by the public and private sector for cultivation under secure conditions is as follows:

Table 15.10: Selection of varieties

Public Sector varieties	:	Pant Parthenocarpic Cucumber-2, Pant Parthenocarpic Cucumber-3 (G.B. Pant University of Agriculture and Technology, Pantnagar), Pusa Seedless Cucumber-6 (IARI, New Delhi), KPCH-1 (KAU, Thrissur)
Private Sector varieties	:	KUK-9, 24, 29, 35; Kian, Hilton, Valleystar; Multistar; RS- 03602833, Kafka, Oscar, Dinamik etc.

Time of sowing and method

Cucumber can be grown successfully in greenhouses during the year. While seed is usually sown directly into the soil, looking at the high cost of the seed and the issue of competition between the plants during the filling of the gap. In total, 3000 seeds are sufficient for a greenhouse of 1000 m². It is advisable to increase 20 % of the overall population by way of plug trays so that they can be used for filling holes in a timely manner in order to keep up with the development of other plants. There are three components, cocopeat, vermiculite and perlite, which are used as a nursery growing tool. These ingredients are combined in a ratio of 3:1:1 before filling the trays. Cocopeat alone can be used as rooting media due to the cost of these materials. Cocopeat typically comes in bricks of 5 kg, but before using it as a rising medium, it must undergo multiple hydration processes with water to extract excess salt present in it. Next phase is to hydrate the calcium nitrate cocopeate brick at a rate of 100 g per brick for at least 24 hours.

Fertilizer management

The fertilization schedule for the production of capsicum in the greenhouse is as follows: **9.0:7.5:7.5 kg NPK/ 1000 m²**

Table 15.11: Fertilizer application schedule

Crop Duration	Distribution pattern / ratio of fertilizers			Remarks
	N	P	K	
First Growth Period (Up to 30 days)	2	3	1	• Fertigation must begin at the emergence of the second-true leaf point.
Second Growth Period (30-60 days)	1	2	3	• Fertigation should be done twice a week.
Third Growth Period (60-90 days)	1	2	3	

If the drippers are at a distance of 30 cm with water Discharge Rate of 2 lph, implement the following irrigation plan for better results.

It is also advised to add 0.5 kg Trichoderma viride, 0.5 l Pseudomonas fluorescens, 2.0 t FYM or 0.4 t vermicompost and 0.5 kg micronutrients (Grade-5) at the time of planting.

Irrigation management

If the drippers are at a distance of 30 cm with water Discharge Rate of 2 lph, implement the following irrigation plan for better results.

Table 15.12: Irrigation schedule

Crop Stage	Time of operation of drip system /irrigation (minutes)	Frequency of Irrigation		
		Summer	Winter	Rainy
Upto initiation of flowering	25	daily	Alternate Day	Every 4th Day
Fruit Setting to First Harvesting	40	Frequency of Irrigation	Alternate Day	Every 4th Day
first Harvest to one week prior to last harvest	35	Frequency of Irrigation	Alternate Day	Every 4th Day

Training and Pruning

Cucumber plants in greenhouse are training to single stem system, which can be achieved by removing all other laterals arising from the axials of leaves, commonly known as suckers at the attainment of 10-12 cm length and only main stem should be allowed to grow vertically along the supporting string.

Fruit thinning

Overbearing can be a problem sometimes. To keep plants from being stressed and to increase the size of the fruit, regulate the amount of fruit per plant by selective fruit thinning. This technique is strong, so use it with caution. The optimum number of fruits per plant varies from the cultivar and even further from the growing conditions. While limiting the number of fruits per plant results inevitably in high-priced big fruits, growers risk underestimating the ability of the crop or failing to predict good weather. They can decide to extract too many fruits and therefore restrict production in an excessive way. Fruit thinning is definitely most effective in the hands of experienced farmers who can use it to optimize their financial returns. The fruit to be pruned must be collected as quickly as possible before it grows too growing. This method involves a great deal of expertise in figuring out the exact crop load in cucumber, and thus, in the presence of an exact number of fruits to be diluted, the common principle of "Survival of the Fittest" can be permitted to survive. Though in the aftermath of this phenomenon, the chances of being deformed and under-sized fruits are greatly improved.

Pest and Disease management

Pest management

» **Leaf Eating Caterpillar** *(Dipahania (=Margaronia) indica* **Saund)**

Management

» Soon after germination, add neem cake to soil.
» Spray any contact insecticides such as Carbaryl 50 WP @ 3g / l. Neem or pongamia soap @ 0.75 percent also controls this insect effectively.
» Soil application of neem cake (once directly after germination and again at flowering) accompanied by NSPE @ 4 per cent and neem 1 per cent alternately at 10-15 days interval.
» Spray Carbaryl 50 WP at 3g / l or Indoxacarb 0.5 ml / l.

» **Serpentine Leaf Miner** *(Liriomyza trifolii* **Burgess)**

Management

- » Soil application of neem cake @ 250 kg / ha immediately after germination.

- » Destroy the leaves of cotyledon with leaf mining 7 days after germination.

- » Spray PNSPE @ 4 percent or neem 1 percent or neem formulation with 10000 ppm or more (2ml / l) after 15 days of sowing and repeat after 15 days, if necessary.

- » If the occurrence is high, first extract all seriously polluted leaves and kill them. Then blend the neem soap with 5 gm and the hostothion with 1 ml / l and spray. Spray neem soap 1 per cent or PNSPE or neem mixture 10,000 ppm or more (2ml / l) after one week.

- » Never apply the same insecticide over and over again.

» **Red Spider Mite** *(Tetranychus neocaledonicus* **Andre)**

Management

- » Spray neem or pongamia soap at 1% on the lower surface completely.

- » Alternatively, spray Dimethoate 30 EC @ 2ml / l or Ethion 50 EC @ 1ml / l or Wettable Sulfur 80 WP @ 3g / l.

» **Thrips** *(Thrips palmi Karny)*

Management

- » Soil application of neem cake (once directly after germination and again at flowering) accompanied by NSPE @ 4 per cent and neem 1 per cent alternately at 10-15 days interval.

- » Spray some systemic insecticides such as Acephate 75 SP @ lg / l or Dimethoate 30 EC @ 2 ml / l.

» **Root-knot Nematodes** *(Meloidogyne incognita)*

Management

- » Seed treatment with bio-pesticide Pseudomonas fluorescens @ 10g / kg of seed.

» Apply 3 G Carbofuran @ 1 kg ai / ha to sowing and repeat after 45 days.

» Apply 2 T of FYM enriched with Pochonia chlamydosporia and Paecilomyces lilacinus per acre before sowing along with 100-200 kg of neem or pongamia cake.

Disease management

» **Anthracnose (*Colletotrichum orbiculare* & *C. lagenarium*)**

Management

» Seed should always be harvested from good fruits and disease-free zones.

» Seeds must be handled with carbendazim at 0.25 per cent.

» Farm sanitation through destroying field waste.

» Grow the crop on the bower system to avoid contact with the soil.

» Maintain proper drainage in the fields

» Seed production would ideally be carried out during the summer season since the summer crop is always free of pathogens.

» Carbendazim foliar spray @0.1 per cent or chlorothalonil @0.2 per cent but spray must be begun immediately after infection.

» **Downy Mildew *(Pseudoperonospora cubensis)***

Management

» Crop should be planted with a wide spacing in well-drained soil.

» Air circulation and access to sunshine tends to monitor the occurrence and progression of the disease. Bower 's crop system decreases the occurrence of disease.

» Field sanitation by burning seed debris to minimize inoculation.

» Seed processing would ideally be carried out throughout the summer season since the summer crop is always disease-free.

» Using accommodating lines of cucumber, such as Summer Prolific.

» Mancozeb Protective Spray @ 0.25 percent at 7-day period provides strong power.

» One spray of Metalaxyl + Mancozeb @ 0.2 percent can be issued in extreme situations, but should not be repeated.

» **Powdery Mildew** *(Sphaerotheca fuligena* and *Erysiphe cichoracearum)*

Management

» Foliar sprays of Penconazole @ 0.05% or Tridemorph @ 0.1% or Carbendazim @ 0.1%, give very good control of the disease.

» Use tolerant line.

» **Fruit Rots** *(Phytophthora cinnamomi, Pythium, Rhizoctonia, Phomopsis cucurbitae)*

Management

» Prevent contact of the soil with the fruit by means of a bower method of cultivating and planting. Have adequate drainage throughout the region.

» Green manure followed by soil application of *Trichoderma* @ 5 kg / ha in soil is very effective in controlling most of the fruit rotting.

» Extract the infected fruit and burn it to high the main inoculum.

» **Gummy Stem Blight** *(Didymella bryoniae*-teleomorph *and Phoma cucurbitacearum* **anamorph)**

Management

» Avoid exotic hybrids and varieties due to high susceptibility.

» Summer ploughing and green manure accompanied by Trichoderma use.

» Ensure adequate ground irrigation and aeration.

» Carbendazim seed treatment @ 0.25 per cent.

» One carbendazim drain @ 0.1 percent near the collar zone.

» Prevent injuries in the collar zone.

» **Leaf Spots** *(Cercospora citrullina, Alternaria cucumerina* and *Corynespora melonis, Didymella bryoniae* (teleomorph) and *Phoma cucurbitacearum* anamorph)

Management

» Soil sanitation, planting of safe seeds and crop rotation minimize the occurrence of disease.

» Fungicidal Mancozeb spray @0.25 per cent alternated with one Hexaconazole spray @0.05 per cent.

» Seed production should preferably be carried out during the summer season because the summer crop is often disease-free.

Mosaic and Leaf Distortion

Management

» Disease control includes the removal of diseased species and plants. Virus free seeds must be used to test the transmission of seeds.

» Initial roughing of contaminated seeds.

» Periodic application of systematic insecticides up to the flowering level to monitor vectors. Seed development will ideally be carried out throughout the summer season, because summer crops are mostly safe from virus infection.

» Restricted usage of rare blends and combinations in bottles of gourd, sour gourd and cucumber.

Yield

Cucumber is usually ready for first harvest in 35 to 40 days of planting, depending on climatic conditions and methods of crop management. Harvesting is achieved when the fruit is more or less cylindrical and tightly packaged and may be completed early in the morning or late in the evening. The substance should be automatically moved to a clear, shaded and ventilated area. When the fruits are harvested manually, they should be cut or snapped with a slight twisting motion and should not be separated from the vines to remove 'pulled ends.' A yield of 10-15 t/100 m^2 can be obtained from greenhouse cucumber.

Production Technology for Melons Under Greenhouse

Scientific name	:	*Cucumis melo*
Common name	:	Sweet melons
Origin	:	Not known

Scientific name	:	*Citrullus lanatus*
Common name	:	Tarbuj
Origin	:	West Africa

The origin of the muskmelon is not identified. Evidence has shown that seeds and rootstocks are among the commodities exchanged along the caravan roads of the Ancient World. Many botanists find muskmelon to be native to the Levant and Egypt, although others find muskmelon to be native to India or Central Asia. Others also favour African origin, and wild musk melons can still be found in some African countries in modern times. Musk melon and water melon are essential vegetables grown over a wide area in various parts of India and are high in demand during hot months. They are dessert fruit mainly eaten for sweetness, good taste and thirst-quenching ability. The typical melon crop is grown by seed sowing from the end of February to the second week of March and is ready for harvesting in May-June. Nevertheless, the crop may be grown in greenhouses for off-season growth in order to achieve very high export rates.

Climate and Soil requirement

Melons can be cultivated in hot and dry conditions, but plants are susceptible to low temperatures and frosts. A humid climate can promote the production of foliar diseases. High humidity and excess moisture at fruit maturity can affect the quality of the fruit. The ideal temperature for plant growth is 28-30°C. Comparatively, low humidity and high day temperature during the ripening cycle, with sufficient sunlight, lead to the production of the flavour and total soluble solids (TSS) in the fruit. Such factors are also ideal for reducing the incidence of foliar diseases.

A well-drained sandy loam soil is ideally suited for melon growing. Melons are slightly tolerant of soil acidity and favor a soil pH of 6.0 to 7.0.

Table 15.13: List of the open pollinated varieties identified/released in India by public Sector

Crop	National level	State level
Musk melon	Kashi Madhu, Pusa Sarbati, Hara Madhu, Pusa, MHY-5, Madhuras, Arka Jeet, Arka Rajhans, Durgapura Madhu, NDM-15, Pusa Rasraj (F_1)	Punjab Sunehari, Punjab Rasila, Arka Rajhans, Hisar Madhur, RM-43, MHY-3, RM-50, Kashi Madhu **Hybrids:** Punjab Hybrid-1, MHY-3, MHL-10, DMH-4
Water melon	Durgapura Meetha, Sugar Baby, Arka Manik, Arka Jyoti (F_1)	Durgapura Kesar, Durgapura Lal RHRWH-12 (F_1)

Table 15.14: Selection of variety

Musk melon	Water melon
Cut fruit varieties are mainly cultivated under greenhouse conditions in Israel and other countries for off-season production and sale to high markets.	Varieties of medium fruit size are favoured.
Many varieties carry andromonoecious flowers except Pusa Rasraj (Monoecious flower) and need artificial pollination for the fruit collection.	Both varieties bear monoecious flowers, which also require artificial pollination for the fruit collection.

Table 15.15: List of cultivars/ hybrids identified by Private Sector for general cultivation

Crop	Name of cultivars/ hybrids
Musk melon	NS-915, NS-89, NS-931, NS-972, Dipti, Madhuras, Madhurima, Urvashi, Madhulika, MHC-5, MHC-6, DMH-4, Bobby,
Water melon	Nutan, Madhubala, MAdhuri, Aashtha, Mithasnina, Khushboo, NS-701, NS-702, NS-705, NS-200, Madhu, Milan Nath- 101, MHW- 4, 5, 6, 11, Mohini, Amrit, Hanoey, Suman- 235, Netravati

Planting

Under greenhouse conditions, melons are typically planted as off-season crops.

Melon seeds may be grown directly on the fields. Nonetheless, in the case of more expensive plants, first of all, the seedlings will be pro-trays or polyethylene in protected conditions. Plants will be eligible for transplantation in 28-32 days. The melons are planted inn-paired rows on each bed in plant 60-45 x 50 cm.

Crop	Seed Rate (per 1000 m²)
Musk melon	100-150 g
Water melon	200-250 g

Training and Pruning

Melon plants are trained upwards so that the main stem of the plants can ascend the overhead wires together with a polyethylene twine. The end of the twine is firmly attached to the base of the stem with a non-slip noose and the plants are tightly vertically to the twine. When the plant hits the horizontal support wires, it can be built together with the steel wire (running on the length of the rows at a height of 8-9 feet) and then the plants are rained down.

Removal of secondary shoots up to the seventh node is found to be effective in musk melon to enhance plant growth and fruit set and to promote early flowering.

The side branches are pruned in watermelon up to 45-60 cm above the bed level. Then the side branches are pruned only after leaving one or two fruit buds (female flowers) which bear fruit. Plants are deliberately trellised, with no damage to the side branches and the female flower on the main stem.

Table 15.16: Use of Plant growth regulators (PGR) for early appearance of female flowers

Crop	Stage of Application	PGR	Dose
Musk melon	First spray at 2-true leaf stage second at 4-true leaf stage	Ethrel	250 ppm
Water melon	First spray at 2-true leaf stage second at 4-true leaf stage	2,3,5- Triiodobenzoic Acid (TIBA)	25-50 ppm

Pollination

Melons require artificial pollination to be properly set because of their flower structure, i.e. andromonoecious in melon musk and monoecious in melon water. Honeybees are the perfect pollinators for greenhouse melon producing crops. One colony of honey bees (*Apis melifera*) with 20,000 bees is adequate for successful pollination in 1,000 m² of greenhouse area. The beehive direction and good ventilation are critical considerations for the effective working of the bees. Honey bees cannot respond well to the greenhouse environment at times, so that hand pollination may also be used to influence artificial pollination. Since the stigma in both crops is receptive early in the morning, the use of hand pollination at peak hours of stigma receptivity must be attempted. Maximum of 3 fruits per vine are kept in musk melon while in water melon plants only 2 fruits are produced.

Fruit support

Small, mesh sacks (onion sacks), cheesecloths or nylon can be used as slings to protect the fruit. The bags may be tied to the trellis or the support cable. The bag is built to allow light penetration and not retain moisture. The bag should not be removed from the trellis until the fruit is ripe. Micro seedless (or seedless) watermelons may be milled in a high tube. When this is the case, the fruit must be sponsored. Other types of melons (large, seeded or seedless) may be grown without a trellis and left to the vine in the high tunnel.

Fertigation

The fertilization pattern for melon production in the greenhouse is as follows:

25:20:30 kg NPK/ 1000 m²

Table 15.17: Fertilizer application schedule

Crop Duration	Distribution pattern/ ratio of fertilizers			Remarks
	N	P	K	
First Growth Period (Up to 30 days)	2	3	1	Fertigation should be started at the appearance of 2nd-true leaf stage. Fertigation should be carried out twice a week.
Second Growth Period (30-60 days)	1	2	2	
Third Growth Period (60-90 days)	1	1	3	

It is also recommended that 0.4 t of vermicompost and 5 kg of micronutrients (Grade-5) be used at the time of planting.

Irrigation management

If the drippers are at a distance of 30 cm with water Discharge Level of 2 lph, implement the following irrigation plan for improved results.

Table 15.17: Irrigaton schedule for musk melon:

Crop Stage	Time of operation of drip system /irrigation (minutes)	Frequency of Irrigation
Upto 30 days	25	Alternate Day
31 to 60 days	40	Alternate Day
After 60 days	35	Alternate Day

Table 15.18: Irrigaton schedule for water melon:

Crop Stage	Time of operation of drip system /irrigation (minutes)	Frequency of Irrigation
Upto initiation of flowering	30	Alternate Day
Fruit Setting to First Harvesting	50	Alternate Day
First Harvest to one week prior to last harvest	40	Alternate Day

Physiological disorders in musk melon:

a. **Fruit cracking:** Fruit cracking of musk melon due to boron deficiency should be observed. Spraying with Boron @ 50 g/25 l water can be undertaken to avoid the cracking of the berries. Fertigation of 50 g of boron can also be done at weekly intervals.

b. **Fruit drop:** Fruit drops is caused by improper pollination. Care should be taken during the method of hand pollination.

Physiological disorder in water melon:

1) Blossom End Rot (BER):

Blossom-End Rot (BER) is a physiological or non-parasitic condition that is related to calcium deficiency, moisture stress or both. Prevention guidelines include proper levels of calcium, appropriate soil pH (6 to 6.5) and a reliable and ample source of moisture. The occurrence of BER is typically very unpredictable from season to season and appears to occur more frequently in oblong melons. Water melons with BER are considered non-marketable.

2) Hollow heart and white heart:

HH&WH are two physiological illnesses that are affected by genetics, environment and potentially a variety of nutritional influences. Only cultivars that have not shown abnormally high incidences of HH or WH should be planted to minimize the occurrence of these two problems. In addition, the crop would be cultivated under optimal (as near as possible) nutritional and moisture conditions. HH and WH impact the quality of melon juice, which may be serious enough to cause potential buyers to reject melons.

3) Sunscald:

Sunscald is a melon damage caused by strong sunlight. Sunscald can be especially serious in dark colored melons. Developing and maintaining sufficient canopy cover to provide protection (shading) for melons can prevent sunscald. Sunscald decreases consistency by making melons less desirable and can predispose melons to rot.

4) Stem Splitting:

Stem Splitting may occur in seedlings grown for transplantation. This concern tends to be associated with high humidity and dehydration that can arise under greenhouse conditions, i.e. watering uniformly to retain soil moisture, preventing wet-dry media cycles and proper air ventilation can help to alleviate these issues.

Pest and Disease Management

Table 15.19: Pest and disease for melons

Insect Pests	Management Practices
Leaf Eating Caterpillars	• Soil application of neem cake (once directly after germination and again at flowering) accompanied by NSPE @ 4 per cent and neem 1 per cent alternately at 10-15 days interval. • Spray Indoxacarb 0.5 ml/liter.
Serpentine Leaf Miner (*Liriomyza trifolii* Burgess)	• Soil application of neem cake @ 250 kg / ha directly after germination. • Remove cotyledon leaves with leaf mining 7 days after germination. • Spray PNSPE @ 4 percent or neem 1 percent or neem mixture with 10,000 ppm or more (2ml / 1) after 15 days of sowing and repeat after 15 days, if required. • If the occurrence is high, then extract any severely infected leaves and kill them. Mix the neem soap 5 gm and the hostothion 1 ml / l and mist. Spray neem soap 1 percent or PNSPE or neem mixture 10,000 ppm or more (2ml / l) after one week. • Never constantly apply the same insecticide.
Red Spider Mite (*Tetranychus neocaledonicus* Andre)	• Spray neem or pongamia soap at 1% on the lower surface completely. • Instead, spray Spiromesifen @1ml / l or Fenazaquin @3g / l.
Thrips (*Thrips palmi Karny*)	• Soil application of neem cake (once directly after germination and again at flowering) accompanied by NSPE @ 4 per cent and neem 1 per cent alternately at 10-15 days interval. • Spray some systemic insecticides such as Acephate 75 SP @ lg / l or Dimethoate 30 EC @ 2 ml / l.

Root-knot Nematodes (*Meloidogyne incognita*)	• Treatment of seed with bio-pesticide Pseudomonas fluorescens @ 10g / kg of seed. • Apply Carbofuran 3 G @ 1 kg ai / ha to sowing and repeat after 45 days. • Add 2 tons of FYM enriched with Pochonia chlamydosporia and Paecilomyces lilacinus per acre before sowing, along with 100-200 kg of neem or pongamia cake.
Diseases management	
Anthracnose (*Colletotrichum orbiculare & C. lagenarium*)	• Seed must be collected from healthy fruits and disease-free zones. • Seeds must be handled with carbendazim at 0.25 per cent. • Field sanitation through burning plant debris. • Cultivate the crop on the bower device to prevent interaction with the dirt. • Ensure adequate land irrigation. • Seed production would ideally be carried out during the summer season since the summer crop is always free of pathogens. • Carbendazim foliar spray @0.1 per cent or chlorothalonil @0.2 per cent but spray must be begun immediately after infection.
Downy Mildew (*Pseudoperonospora cubensis*)	• Crop should be planted with a wide spacing in well-drained soil. • Air circulation and access to sunshine tends to monitor the occurrence and progression of the disease. • Bower's seed scheme decreases the occurrence of disease. • Field sanitation by burning plant debris to minimize inoculation. • Seed processing would ideally be carried out throughout the summer season since the summer crop is always disease-free. • Mancozeb Safe Spray @ 0.25 percent at 7-day period provides strong power.

	• One spray of Metalaxyl + Mancozeb @ 0.2 percent can be issued in extreme situations, but should not be repeated.
Powdery Mildew (*Sphaerotheca fuligena and Erysiphe cichoracearum*)	• Penconazole foliar spray @0.05 per cent or Tridemorph @0.1 per cent or Carbendazim @0.1 per cent provides very strong control of the disease. • Using the resistance rows.
Gummy Stem Blight (*Didymella bryoniae* teleomorph *and Phoma cucurbitacearum* anamorph)	• Remove rare combinations and variations owing to increased sensitivity. • Summer ploughing and green manure accompanied by Trichoderma use. • Ensure adequate ground irrigation and aeration. • Carbendazim seed treatment @ 0.25 per cent. • One carbendazim drenching @ 0.1 percent near the collar region. • avoid injuries in the collar zone.
Leaf Spots (*Cercospora citrullina, Alternaria cucumerina and Corynespora melonis, Didymella bryoniae* (teleomorph) and *Phoma cucurbitacearum* anamorph)	• Soil sanitation, selection of healthy seeds and crop rotation reduce the incidence of disease. • Fungicidal Mancozeb spray @0.25 per cent alternated with one Hexaconazole spray @ 0.05 per cent. • Seed processing would ideally be carried out throughout the summer season since the summer crop is always disease-free.
Mosaic and Leaf Distortion	• Disease control includes the removal of diseased species and plants. • Virus free seeds must be used to check the transmission of seeds. • Initial rouging of infected plants. • Periodic application of systematic insecticides up to the flowering stage to monitor vectors.

Harvesting Indices

Table 15.20: Harvesting stage of melons is given as under:

Crop	Days to first harvest	Harvesting indices/stage	Yield per 1000 m²
Musk melon	75-80	When the fruits slip easily from the vine. Fruit should show changes in color and degree of netting, and a softening at the blossom end. Best eating maturity follows in one to three days; and best flavor is attained if musk melons are held near 21 °C for this final ripening then chilled for serving.	6-8 ton
Water melon	90-110	Fruits are harvested when they achieve full size. At this point, the curly tendril closest to the fruit attach is always shriveled or dry.	18-20 ton

Production Technology of Cabbage

Scientific name	:	*Brassica oleracea* L.
Common name	:	Band gobhi
Origin	:	Western Europe

Cabbage is a typical cool-season crop grown with a thickened central bud called the head. It is one of the most common and widely grown vegetables in the area and was second to the production of potatoes. This is a rich source of vitamin A, C and minerals, including potassium, calcium, sodium and iron.

Soil and Climate

Early crop light soils are best, whereas late crop heavy soils are favoured. The optimal pH of the soil is 6.0 to 6.5. This needs a cold and humid climate. Across the lower hills of Meghalaya, cabbage is mainly grown during the winter season (October-January), while in the higher hills it is grown during the rainy and winter seasons. Throughout the middle of the hills, cabbage can be grown nearly all year round.

Varieties:

Pusa Ageti, Green Express, Green Challenger, Green Hero, Pride of India, Rare Ball

Field preparation

For field preparation, the soil is ploughed 2-3 times using a power tiller or a spade. Planking is performed after the last ploughing to create a friable soil layer for transplantation. The raised beds are 1 m wide, 4-5 m long and 30 cm above the ground.

Time of Sowing

Early season: Mid June to July,

Mid season: Mid August to September,

Late season: October - November

Seed Rate

For early season 500g/ha, while during the mid & late season 400g/ha required.

Nursery Raising

The nursery bed should be prepared widely by applying well rotten FYM or compost @ 4kg / m². Before sowing, the seeds should be handled with Captan or Thiram @ 2.5 g / kg of seed to get rid of fungal diseases. The seeds are planted at a spread of 2-3 cm between the plants and 8-10 cm between the leaves. The range of sowing is 1-1.5 cm. After sowing, the seeds are covered with a mixture of fine soil and sieved FYM. A light irrigation should then be applied. The nursery bed must be kept safe from weeds.

Transplanting

5-6 week old seedlings with 4-6 leaves should be transplanted. Transplanting should be done in the evening. Immediately after transplanting, irrigation should be applied.

Spacing

45 x 45 cm should be kept for early season spacing, while 60 x 45 cm should be kept for mid and late season.

Manure and fertilizers

FYM or compost @ 15 to 20 tons / ha is incorporated into the soil during the preparation of the land. Including FYM, N: P: K @ 120:60:60 kg / ha is used. A total amount of phosphorus and potash, along with half the amount of nitrogen, is used at the time of transplantation. The remaining dose of nitrogen is delivered as a top dressing at two separated levels, i.e. 30 and 45 days after transplantation.

Weeding and earthing up

For the entire lifespan of the crop, two or three weeding are important to control the weeds followed by earthing up.

Pest and Disease Management

Pest management

» **Cutworms:** The caterpillars are 3 to 4 cm long, grey or brown or almost black with various markings. They're hiding in the morning and feeding in the night. They do damage by chewing the leaves and cutting off the young seedlings just above the ground level.

Management

1. Increasing of paired rows of mustard per 25 rows of the field.

2. The collection and destruction of the larvae at the early stage of the crop.

3. Application of a furadan to crop which heavily infested

» **Diamond backmoth**

Management

» Growing of mustard as intercrop as a ratio of 20:1 to attract diamond back moths for oviposition. Spray the mustard crop with insecticide to avoid the dispersal of the larvae periodically.

» Spray the NSKE 5 % after primordial stage.

» Spray Cartap hydrochloride 1 g/lit or *Bacillus thuringiensis* 2 g/lit at the primordial stage (ETL 2 larvae/plant)

» Release of parasite *Diadegma semiclausum* at 50,000/ha, 60 days after planting.

» Install the pheromone traps at 12 No. /ha.

» **Aphids**

Management

» Spray neem oil 3 % with 0.5 ml Teepol/lit.

» Install yellow sticky trap @12 no. /ha to monitor "macropterous" adults (winged adult).

» **Leaf Webber:** The leaves are skeletonized by the larvae which remain on the underside of the leaves in the webs and feed on them. They strike flower buds and seeds, too. The bug is usually used to suck early-grown crop.

Management

» The processing and killing of the larvae at the early stage of the crop.

» Cyfluthrin @ 0.5mll of water will be used to spray the crop.

Disease management

» **Damping off**: It's a serious disease in the nursery. In extreme circumstances, the infected seedlings droop and fall due to contamination in the collar area.

Management

» Treatment of seed with Thiram or Captan @ 2.5-3 g / kg of seed. Seedlings should be treated with Hexaconazole 5 per cent + Captan 70 per cent WP or Metalaxyl-M + 640 g / kg Mancozeb @ 2g / l of water.

» **Black rot:** The first symptoms of the disease frequently occur around the edges of the leaves as chlorotic regions and chlorosis progresses into the middle rib creating a V-shaped layer. Symptoms growing occur on either side and in the middle of the leaves. The bacteria are spread via the crop. The use of black rot-tolerant varieties is the safest way of disease prevention.

» To encourage drainage, planting on raised beds should be carried out.

» Major reductions in disease have been reported as seeds are treated with Agrimycin-100 (100ppm) or Streptocycline (100ppm).

» Plants should be closely inspected for symptoms of black rot, and contaminated plants should be removed and burned.

Harvesting and yield

Harvest is accomplished when the heads are well-developed and firm. The heads are separated with a knife, usually with all the wrapper leaves. A good crop may yield between 250 – 300 kg / ha.

Production Technology of Gerbera under Greenhouse

Scientific name : *Gerbera jamesonii*

Common name : African daisy

Origin : Tropical region of South America, Africa and Asia

Gerbera is commonly referred to as Transvaal daisy, Bar berton daisy or African daisy. It's an important commercial cut flower crop. Gerbera flowers have a wide range of colours, including yellow, orange, cream-white, pink, brick red, red, terracotta and various other intermediate colours. Bicolor flowers are also available in double varieties. Gerbera flower stalks are long, thin and leafy, with a long vase life. It flowers all year round in warm, humid conditions. Gerbera can also be cultivated as an open-air field crop on raised beds, as a greenhouse plant under controlled conditions (protected cultivation) and as a potted plant. There are other purposes of gerbera. It is grown as a landscape plant for beautification or on flower beds, borders and in a rock garden. Flower arrangements are also appropriate in a vase. This is very popular to use in a flower bouquet. Gerberas may be planted with seeds, buds split with clumps and tissue culture. The largest manufacturing counties in India are Karnataka, Maharashtra, Tamilnadu, West Bengal, Himachal Pradesh, Jammu & Kashmir and Gujarat.

Soil

There are two main considerations to be addressed in the collection of soil for the production of Gerbera.

1. The pH of the soil should be between 5.5-6.5.

2. The soil salinity standard shall not surpass 1 ms / cm;

3. For better root growth and better root penetration, the soil should be highly porous and well drained.

Red lateritic soils are good for the cultivation of Gerbera, as they have all the essential characteristics that the ideal soil should have.

Climate

Bright sunlight promotes the development and quality of the flowers, but this flower prefers indirect sunlight in the summer. Gerbera plants raised in low-light environments do not bloom well. The maximum day and night temperatures are 27°C and 14°C respectively. The optimal temperature for flower initiation is 23°C and for leaf unfolding it is 25-27°C.

Varieties

Important cultivars of Gerbera : **Dalma (white), Savannah (red), Rosalyn (pink), Cream Clementine (cream white), Dana Ellen (yellow) and Maron Clementine (orange).** Alexias, Anneke, Balance, Cacharelle, Diana, Rosetta, Jaffa, Pre Intenzz, Sangria, Thalsa, Pricilla,Sonsara, Winter Queen, Paganini, Nette, Gloria, Ginna, Ingrid, Intense, Sunway, Stanza, rosaline, Zingaro, dune and Monique.

Preparation of planting bed

In general, Gerberas are developed on raised beds to assist with easy movement and improved drainage. The scale of the bed will be as follows:

» Height of bed: 1.5 feet (45 cm)

» Bed width: 2 feet (60 cm)

» Between the beds: 1 feet (30 cm)

The raised beds are 80 cm wide and are suitable for planting. The base should be 90 cm and the height should be 45-60 cm. Gravels can be placed at the bottom of the bed for better drainage. Leave 45 cm of space between the beds as a working space. Attach and combine 2.5 kg of single super phosphate and 0.5 kg of magnesium sulfate per 10 sq.m. Bed area. The beds for planting should be rather moist, well drained and airy. Gravel / sand may be applied to the bottom for improved drainage. Organic manure is recommended to improve soil texture and nourish gradually. The soil should be lose all the times. Organic manure and soil should be thoroughly mixed for optimum results. After irrigation, the soil should not be very compact. The upper layer of soil and FYM should be correctly mixed together. When preparing the bed, add single super phosphate (0:16:0) @ 2.5 kg per 100 sqft for improved root establishment and magnesium sulfate @ 0.5 kg per 100 sq.ft. To take care of Mg's deficit. Neem cake (@1 kg / m) is often applied to avoid nematode infestation.

Soil sterilization

Sterilization of soil is necessary to reduce the risk of *Phytophthora*, *Pythium*, *Fusarium*. Formalin or formaldehyde (1 liter in 10 liters of water) is sprinkled (1-2 liters / sqm) with rose can on a mixed growing medium layer of 1-1.5 feet in height and covered with plastic. After 1 week, remove the plastic. Flush the remains of formalin with water of 50-80 lit / sq. m. Wait for a week for a strong soil tillage state. There are three major methods of soil sterilization available.

1. ***Steam*** : Not applicable to Indian circumstances.

2. ***Solar :*** In this process, plastic sheets are covered on soil for 6-8 weeks. The sunlight heats the soil, so it destroys much of the fungus.

3. ***Chemical :*** This is the most advanced and most effective method. Hydrogen peroxide (H_2O_2) with silver is used for soil sterilization. The use of formalin @ 7.5-10 lit/100sqm can also be done.

Hydrogen peroxide (H_2O_2) with silver

Process:

1. Wet the beds with the irrigation.

2. Mix water with hydrogen peroxide at a rate of 35 ml per litre.

Add this remedy uniformly to soil beds. Using one liter of mixture solution (H_2O_2 peroxide with silver + water) per one meter area. Following that, the crop can be planted in 4 to 6 hours.

Benefits of Hydrogen Peroxide (H_2O_2) with Silver:

1. Economical, reduce input costs

2. This method, which is very easy and safe, has no harmful effect on human health.

3. Planting can take place after 4 to 6 hours of fumigation.

4. Eco-friendly and does not contain phytotoxic effect on plants.

5. Almost all fungi, bacterial and virus activity in the environment kills larvae and insect eggs.

Propagation: Gerbera is propagated by plants, through cuttings of side shoots and suckers.

Seeds: Seed is placed when cross-pollinated. Seed sowing can be achieved in almost any season. Seeds germinate at 15 to 20oC within two weeks; otherwise they can take up to 30 days. The seedlings will blossom in the second year and will produce healthy flowers from the third year onwards.

Vegetative: Side shoots, with a certain amount of heel, are used. Divisions / suckers, cutters are often required.

Micropropagation: The plant sections used as micropropagation explants are shoot tips, leaf mid-rib, capitulum, flower heads, inflorescence and buds. Murashige and Skoog (MS) media with alteration was effectively used as culture media.

Planting

Plant should be no less than three months aged. The plant should have at least 4 to 5 leaves at the time of planting the tissue culture. Gerberas are placed on a raised bed in the shape of two rows. The Zigzag plantation method is much favoured. When planting 65 per cent of the root ball should be kept below ground and the remainder of the section, i.e. 35 per cent should be kept above ground for improved air circulation in the root areas. Ideal planting density and spacing: 8-10 plants per square meter or 30 x 30 cm or 40 x 25 cm.

Fertilization

Irrigate and also fertilize in limited amounts for optimum results. Often evaluate the soil once every 2-3 months to establish a precise nutrient plan. 25-75 t / ha of well-decomposed organic manure is necessary. For the first three months after planting, application of 20:20:20: N: P: K @ 1.5 g / l of water every two days during the vegetative process stimulates enhanced foliage. Do not include poultry or other manures. Soil and FYM are mixed in a ratio of two to one parts. The addition of rotted manure to the soil prior to sterilization is a must.

Once flowering begins, N: P: K 15:8:35 shall be made available at a rate of 1.5 g / l of water / day. Micronutrients can be delivered regularly or fortnightly, based on signs of deficiency (preferably chelated source). Boron deficiency causes the base of young leaves to turn black. Zinc deficiency symptoms can be identified with the C-shaped leaf structure caused by chlorosis on one half of the leaf blade that ceases to spread, while the other half of the leaf is normal.

Cultural practices

Weeding a raking of soil:

Weeds take the plant nutrients and impact the growth cycle. They should therefore be removed from the bed. Due to daily irrigation, the surface of the gerbera bed is hard and the soil is raking with the help of a raker. Increases soil aeration in the root zone of the plant. This activity will be carried out on a regular basis, and may take place twice a month.

Disbudding:

Removal of poor flowers at the initial stage following planting is termed disbudding. Normal production of gerbera plants begins after 75-90 days from the date of planting. The development of flowers starts 45 days after planting, however the initial production is of poor quality, so such flowers should be withdrawn from the base of the flower stalk. It tends to keep the plant solid and stable.

Removal of old leaves:

Sanitation tends to prevent the illness and insect infestation below the economic level. Old, dried, infested leaves should be collected from the plant and removed from the site of production.

Pest and Disease Management

Pest Management

Thrips, leaf miners and mites are very common pests and need to be removed immediately as and when they appear. The amount of spray solution can be doubled or tripled as the plants grow to cover all plants.

Insect pests: white flies; red spider mites; nematodes; aphids; leaf miners; caterpillars

Insect pests: white flies; red spider mites; nematodes; aphids; leaf miners; caterpillars

Pest management

1. Under protected cultivation circumstances, the usage of insect-proof screens serves as physical obstacles to the elimination of insect pests.

2. Key pest management practices are sanitation with regard to the

use of pest free planting materials, soil solarisation and removal of infested plant parts.

3. Prudent fertilization on the basis of good use of the nutrients to be practiced. Excess Nitrogen should be avoided.

4. The application of carbofuran at 2 kg a.i./ha in combination with neem seed powder @ 100 g / m² is successful for root knot nematode control.

5. Leaf spot gerbera disease may be managed by the care of plants with benomyl (0.1 percent) supplemented by Kavach (0.2 percent).

Disease management

Important Diseases : Root rot (*Pythium irregularae, Rhizoctonia solani*) ; Foot rot (*Phytophthora cryptogea*) ; Blight (*Botrytis cinerea*) ; Powdery mildew (*Erysiphe cichoracearum, Oidium crysiphoides*) ; Sclerotium rot (*Sclerotium rolfsii*) ;Leaf spots (*Phyllosticta gerberae, Alternaria spp.*)

Table 15.21: Viral disease (Cucumber mosaic virus and Tobacco rattle virus)

Disease	Control measures
Root rot	This can be controlled with Captan, Benlate, Aliette drench to soil (2 g/L)
Crown rot	Aliette, Topsin-M (2 g/L)
Powdery mildew	Wettable Sulphur spray (1.5g)
Alternaria leaf spot	Dithane M-45 (Mancozeb) spray (1.5g)

(a) Alternaria leaf spots in gerbera (Dehradun) (b) Damage to gerbera flowers by *Spodoptera litura* in Dehradun (c) Root-knot nematode *Meloidogyne incognita* in gerbera in Jassowala

Harvesting

The first flowers may be harvested after 75 - 90 days after planting. Plants have a productive life up to 24-30 months. One plant produces 35 to 40 flowers per year. it can yield **80 to 100 flowers in span of 30 months.** Flowers of most of the varieties (single types) are ready to be picked when 2 - 3 whirls of stamens have entirely been developed. Some varieties are picked little riper, especially the double types. The good flower has stalk length of 45-55cm, and diameter of the flower is 10 - 12cm.

Morning or evening is best time for gerbera flower harvesting. Skilled labours are required for harvesting of gerbera cut flowers. After harvesting the flowers should be kept in a bucket containing clean water. Flowers are very delicate hence they should be carefully handled otherwise can be damaged and their quality gets deteriorated. For harvesting gerbera no secateurs are required and are done by naked hands.

Post-Harvest Quality and Management

Flowers with a stalk length of 45-55 cm and a diameter of 10-12 cm have a very good price as 'A' flowers. Vase life is 8-10 days. For the best vase life, put the cut end of the flower in clean water at 15 degrees centigrade for 4 hours.

Add 10 ml of sodium hypochlorite to 1 liter of water before flowering. It is possible to use corrugated box / cartons for long-distance transport. Gerbera doesn't need to cool like rose or flesh and has a relatively long shelf life.

Transportation by train and bus is necessary to enter local markets within 12-36 hours. Flowers get the price of Rs. 40-50 / ton that spikes during New Year's season, holidays, Christmas, Valentine's Day festivities, etc. Flowers get a very good price during the wedding season. Growers may use financial facilities under different state and central government schemes to render the enterprise very remunerative.

Production Technology of Carnation Under Greenhouse

Scientific name	:	*Dianthus caryophyllus*
Common name	:	Carnation, Clove pink
Origin	:	Mediterranean region

Carnation is one of the major cut flowers in the world. Carnation are also called clove gilly flowers, divine flowers, gilly flowers and clove pinks. Dianthus caryophyllus is herbaceous perennial plant that grows up to 80 cm tall. The leaves are glaucous greyish green to blue-green, slender, up to 15 cm long. The flowers are grown single or up to five in a cyme; they are around 3–5 cm in diameter and sweetly scented; the original natural flower color is bright pinkish-purple, but cultivars of other colors, including white, red, blue, yellow and green, have been established along with some white colored lines. The fragrant, hermaphrodite flowers have a radial symmetry. The four to six surrounding the calyx, egg-shaped, sting-pointed scales leaves are only ¼ as long as the calyx tube.

Soil and climate

The ideal weather for carnation development should include a calm and steady temperature, low humidity and long days of high sunlight intensity. Tyipical carnation prefers a warmer environment than the spray kind. Plants must be secured from rain and dew during flower growth. Wet plants, especially wet buds and flowers, are susceptible to fungal disease.

Varieties:

Carnation varieties may be demarcated into three major classes based on the size and application of the flower.

» **Standard carnation** - These varieties have single large flower used as cut flower on an individual stem.

» **Spray carnation** - Spray carnation is typically a bunch of flowers with a single stalk on short branches. The flowers on each branch are small, and compact.

» **Micro carnation** - These have shorter stems and more production than spray varieties. Besides its utility in flower arrangement, these are used as ornamental pot plants.

It is best to take up multicolored cultivation in a single carnation polyhouse. There is strong demand for many colours.

New cultivars are being imported and grown in India, produced by different temperate-country commercial companies. When selecting varieties for planting, caution must be taken in addition to colour, about their adaptability to tropical condition and disease resistance. Nearly all new carnation varieties are covered by royalties; hence, growers must be very careful about procurement of the material from the right sources, especially when producing for export purposes.

Light

» Carnations are long day plants that need a high degree of light to grow flowers of high quality.

» The minimum natural light intensity needed for adequate carnation photosynthesis is considered to be about 21500 Lux.

Cultural operations

Polyhouse structures

» To sustain quality and yield, carnation must be grown under protected structures. The greenhouse design should be kept as simple as possible but to maintain the correct temperature in the greenhouse it should meet certain specific demands.

» The plastic greenhouse will have air space of 3 to 4 m³ per m² of floor surface. This relates as a minimum to a gutter height of 2.5 to 3.5 metres.

» If the greenhouse becomes too low, high air humidity and high temperatures can occur which are not ideal for carnation.

» It is also important not to build too large a greenhouse under one roof (maximum 100x50 m = 5000 m² per unit) for good climate control / good ventilation and to make it possible to roll up the sides.

» The roofing material must be 200 micron thick Low Density Polyethylene (LDPE). One kg of this material encompasses 5.37 m². Once in three years LDPE requires replacement.

Propagation

» Carnation (*D. caryophyllus*) for the cut flower is multiplied by cuttings.

» There are annual carnations multiplied by seeds. These annual carnations are suitable for potted ornamental plants.

» *D. Chinensis* and *D. Barbatus* would be multiplied by seeds.

Land preparation

» Deep ploughing to be followed before bed preparation.

» A well-drained porous soil with an adequate organic matter content is needed for good growth.

» As the carnation has a fibrous root system with a sufficient amount of organic matter and microorganism activity in the soil is required to support the growth of a good root system.

» The optimum pH of the soil should be around 6.5.

» The addition of calcium carbonate or dolomite limestone corrects the acid condition and supplies calcium and magnesium for plant nutrition.

» The addition of gypsum or the use of acid forming fertilizers will also reduce the pH of the soil if it is on the higher side.

Bed preparation

» A bed with a width of one meter and a suitable length must be equipped to operate perfectly. The working route is 50 cm. This can be provided between the beds.

» Rising beds are ideal for cultivation.

» The bed may be best prepared by mixing well rotting coir pit / paddy husk / saw dust.

» An appropriate quantity of manure and sand will be applied based on the mineral composition of the site.

» Ordinarily, a well-prepared bed would comprise a mixture of soil, farmland manure, sand and coir pit of 4:2:1:1, respectively.

Spacing and planting

» Planting may be performed in different spaces.

» Usually, 30-45 plants per sq m. This is found to be optimal.

» The common size accompanied by the branching pattern of the plant is 15x8 cm, 15x10 cm, 15x15 cm and 15x20 cm.

» In the case of a shortage of knowledge on the plant pattern 15x20 cm spacing may be observed easily.

» Alternative planting in adjacent rows is advantageous in terms of decreased disease occurrence.

» Shading should be provided for a few days at the beginning of the harvest.

» Care should be taken to maintain humidity in order to prevent plants from drying out.

Staking

» Carnation crop tends to bend unless properly supported. As a consequence, the crop needs support when developing.

» The metal wire woven with nylon mesh is a good supporting material.

Pinching

» Pinching refers to the removal of the rising tip of the main shoot to promote the development of the side shoots.

» Depending on the need for crop spread and consumer demand, single, one and a half pinches and double pinches are given.

» The first pinch is given when the plant reaches six nodes. That is refers to as a single pinch. It will give way to six faces.

» For a 'one-and-a-half pinch,' 2-3 of these lateral shoots are pinched again.

» All the lateral shoots are pinched up for the 'double pinch.'

» Good time for pinching in the morning.

Disbudding

» In carnation, disbudding refers to the elimination of unnecessary buds, such that the remaining bud provides sufficient food for its complete growth.

» In normal carnations, side buds are eliminated where, as in spray carnations, terminal buds are removed.

Nutrient management

» Carnation is highly responsive to the usage of fertilizers. It is very sensitive to boron and calcium deficiency.

» Carnations need nearly equivalent quantities of nitrogen and potash.

» Supplying trace elements of boron and calcium from the medium of development constantly is important for safe flowers.

» For the first three weeks after planting, no fertilizer is needed.

» The RDF used at Hi Tech Horticulture RAU, Pusa is 140:80:120 g/m^2 NPK, of which 50 % of RDF was added as a basal dosage and the remaining 50 % was supplied by fertigation.

» A basal dosage of 70:40:60 gm / m^2 is issued at planting time.

» Remaining fertilizer is added in equally distributed dosage.

Water management

Carnation is very sensitive to the stress of moisture. A small amount of moisture has a major effect on plant growth and the production of branches in the crop. Emergence of branches occurs in flower yield or drop. In spite of this, irrigation was planned on a regular basis to satisfy the necessary water requirements. Overhead sprinklers are needed for the first three weeks to keep young plants from drying out. Drip irrigation should be used later.

Pest and Disease Management

Pest management

» **Thrips** - *Thrips tabaci*

Yellow nymphs and brown colored adults harm developing buds and flowers by sucking the sap.

Symptoms

» Damaged buds have brown lines on light colored flowers and white lines on dark colored flowers. In extreme cases of attack, the flowers become skewed with burnt margins contributing to a significant reduction in quality.

» **Two spotted spider mite** - *Tetranychus urticae*

It is a devastating pest on carnation growing under protected conditions, particularly during the months of September-November and February-May.

» **Bud borer** - *Helicoverpa armigera*

Eggs are laid on growing buds throughout September-October and the damage continues until February-March.

Symptoms

» Young larvae bore inside rising buds and feed on petals which make them hollow. Larvae often strike open flowers resulting in a fall of petals and a significant loss of yield.

Prevention/Control

» Collecting and killing larvae reduces further infestation.

» Spray quinalphos 25 EC @ 2ml / l or methyl parathion 50 EC @ 1 ml / l at intervals of 15 days.

» Spray 14.5 SC indoxacarb @ 1 ml / l or 75 WP thiodicarb @ 1g / l if the occurrence is extreme.

» HaNPV (*Helicoverpa armigera* Nuclear Polyhedrosis Virus) spraying @ 250 LE / ha accompanied by the neem formulations 1-2 ml / l or 40g / l kernel seed extract.

» **Nematodes**

The root knot nematode causes the feeder roots to have small galls that affect the uptake of nutrition and water.

Symptoms

» Foliage goes to show nutritional deficiency such as symptoms accompanied by a decreased flowers size

Management

» To raise nurseries in soil enriched with biopesticides.

» Before planting, apply FYM enriched with bioagent *Pochonia chlamydosporia* and *Paecilomyces lilacinus.*

» Apply Carbofuran 3G at 5 gm per square meter when transplanted and repeat after 45 days.

» Tolerant lines such as IIHRP-1 can be used for cultivation that can withstand nematode infestation.

Disease management

» **Vascular wilt (*F. oxysporum f.sp. dianthi*)**

Vascular wilt is the most severe carnation disease.

Symptoms

» Infected plants first develop a dull green color, gradually wilt and transform straw color. Symptoms sometimes arise in one part of the plant.

» Plants may be targeted at any stage; young plants may unexpectedly dry up, or older plants may grow a pale green color followed by lower leaf wilting.

» This is sometimes accompanied by the general wilting and death of the whole plant.

» In the initial stage of infection, plants can wilt in the middle of the day where the temperature is high and seem to be healing at night. Many plants cannot exhibit symptoms of infection until they begin to bloom, until they unexpectedly fail. When the stem is removed, brown discoloration or streaking can occur in the vascular tissues.

» **Bacterial wilt *(Pseudomonas caryophylli)***

In general, the wilting of plants due to fungi, bacterial wilting triggered by Pseudomonas is often identified in the carnation.

Symptoms

» The main signs of infection include the wilting of plants or shoots and the breaking of stems and yellow lines in the vascular system.

» The tissue immediately below the epidermis becomes sticky and the cutting through the stem reveals discoloration and bacterial odor.

» The leaves grow gray-green, then yellow and dried. Rotting the roots and browning of the vascular tissue are the other symptoms that are commonly seen.

» Cracks form in the internode tissue. Slime oozes out of these cracks while the humidity is high.

» The illness tends to propagate easily at high temperatures.

» **Leaf spots** *(Alternaria dianthi* and *A. dianthicola)*

Symptoms

Pale tan to brown spots with purple borders appear on the leaves. In extreme instances, the leaves become blighted owing to the fusion of the lesions.

Prevention/Control

» Efforts to improve plant vigor by fertilization and irrigation are beneficial.

» Watering should be done early in the day, though, to allow the leaves a chance to dry before the weekend.

» It is important to pick up and extract symptomatic leaves as soon as they grow.

» Infection can be managed through the use of Captafol (0.2 percent) or Chlorothalonil (0.2 percent) or Iprodion formulations (0.2 percent).

» **Foot rot**

Several soil-borne fungi target the root and collar portions of the carnation, especially under high soil moisture conditions. Infections with *Phytophthora* are caused by *P. nicotianae* var. *P. nicotianae* and *P. nicotianae* var. it's infectious.

Symptoms

» Leaves and buds of flowers become infected under wet weather conditions. Rotting the stem at soil level and eventually the infected plants may rot. As the roots become tainted, the leaves slowly discolor and tend to dry from the bottom upwards. Soft rotting of young leaves and buds of flowers are typical signs.

Prevention/Control

» As all of these pathogens survive well under high humidity conditions, the management of moisture is an essential phase in the control strategy.

» The fungicides useful are Benomyl 0.2% or Iprodion formulations 0.2% against *R. solani* and *S. rolfsii*, whereas, metalaxyl + Mancozeb 0.2% and

Fosetyl-Al 0.2% are effective in managing infections by *Pythium* and *Phytophthora*.

» **Flower rot** *(Botrytis cinerea)*

  » The flowers turn brownish color and are coated with gray, fuzzy masses of mycelium fungi. Brown streaks grow on the leaves.

  » This disease becomes especially severe during times of extended cloudy, hot and wet weather.

Prevention/Control

  » Effective sanitation practices, including plant grooming and elimination of spent or senescing flowers, can mitigate infection. It is also important to avoid wetting the flowers while watering the plants.

  » Appropriate field spacing will facilitate good air circulation.

  » Prevention can also be accomplished by the use of fungicide sprays as soon as signs become apparent.

Table 15.22: Disease chart and management

Pathogen/ Cause	Disease	Symptoms	Management
Alternaria dianthicola or *Alternaria dianthi*	Alternaria Leaf Spot	A slight purplish mark on the leaves. Their centers are gray and their leaves become yellow.	Use chlorothalonil (0.2 per cent), mancozeb (0.2 per cent), copper hydroxide (0.2 per cent) or mancozeb (0.2 per cent) + thiophanate methyl (0.2 per cent) to secure safe plants.
Pseudomonas caryophylli	Bacterial Wilt	The leaves turn grey-green, then yellow and dried. Roots rot, vascular tissue browns and cracks form in internode tissue. Slime tastes like these holes while the humidity is high	Placed in pasteurized raised beds. Using safe, disinfected devices. Kill the contaminated vine.

Harvest

Generally, after planting the carnation, it takes 110-120 days to hit the height of flowering. A commercial carnation farm is capable of producing 8 to 20 flowers a year. Depending on consumer demand, the processing of buds with bud nets can be postponed. When buds begin to show colour, bud nets may be positioned covering the whole bud of the flower, which often increases the size of the bud in addition to regulating the flower development.

Post-harvest operations

Immediately after cutting, the cut ends of the flesh flower stems should be put in a bucket containing clean water or a floral preservative containing the biocide and acidifying agent. Such flowers are pre-cooled as early as possible at 20C for two to four hours. Also safe and standard cut flowers are rated on the basis of stem length and bud height. As per consumer requirements, bunches of 10, 15 or 20 flowers are tied with rubber bands and flowers lined with butter paper sleeves. The leaves on 1/3 of the basal stem are cut and the ends of the stems are clipped. Such grouped and cut flowers are quickly placed back in a bucket of preservative solution packed in de-ionized water. Acidic (pH 4.5) with 2-5 percent sucrose and a biocide for pulsing would be a reasonable preservative option for carnations.

Market and trade

The USA becomes the largest producer of carnation in terms of total quantities. Still England is the most eaten per capita. Carnations are also common in Eastern Europe, where roses are almost roses.

Production Technology of Rose Cultivation

                    Scientific name   :   *Rosa spp.*
                    Common name   :   Gulab

According to the widely accepted taxonomy, there are 120 species belonging to the Rosa genus. Species are present in temperate northern climate areas and in subtropical parts of the world. This includes areas from the polar circle to New Mexico, Ethiopia, the Himalayas, Bengal and southern China in the Far East.

Rose is the world's leading cut flower grown commercially. It ranks first in global cut flower trading. The rose not only referred to its position as the King of Flowers, but also as the most popular flower in the world. The high demand for rose-cut flowers in the European markets is mainly due to a lack of local production due to severe winters from November to March. Luckily, this is the most congenial state for the productive growth of most flowers, like roses in India. It is pointed out that the buyer prefers very high quality rose cut flowers on the international market. Since it is difficult to obtain good quality cut flowers under open conditions throughout the year, crops should be cultivated in the greenhouse to produce good quality produce.

Rose is a famous and popular flower; the rose flower is a sign of love all over the world. In Greenhouse, mainly dutch rose grown, the Dutch rose has a strong demand in both national and foreign markets. As a result, Dutch rose cultivation is growing day by day, and the Indian Government is also supporting Dutch rose farming by offering subsidies. Strong sunshine influences the dutch rose output of the plant requiring bright light. The dutch rose therefore grows really well in the Greenhouse, where the atmosphere is under charge.

The Dutch rose was mainly cultivated for export purposes. In India, Dutch roses are exported to countries such as Europe, Singapore, Malaysia, the Middle East, Australia and New Zealand. The Roses plant produces flowers faster in long days (summer) than in short days of winter. This crop provided some shade during the summer season, such as shade net or painting polythene from outside the polyhouse.

Basic essential for Dutch rose to grow

» Light – 40000-60000 LUX
» Temperature 15 – 18 °C
» Humidity 60% -65%
» Good quality water
» Good growing medium

Soil

In order to succeed in Dutch rose cultivation, soil selection is very important. The main factors to be considered are as follows:

» The pH of the soil would be between 5.5-7.
» For improved plant development and stronger root penetration, the soil should be extremely porous and well drained.

Dutch rose also cultivated artificial material, such as coconut peat, rock wool, pumice.

Climate

The ideal temperature range for the cultivation of roses is 15-27⁰C. Most commercial rose cultivars are best grown at 15.5⁰C at night temperature. Approximately 60 to 70% relative humidity may be considered optimal for roses; high humidity, particularly in combination with low temperatures for a long period of time, may trigger problems with diseases such as mildew and botrytis.

Soil sterilization

Before Dutch rose cultivation, soil sterilization is necessary and hydrogen peroxide (H_2O_2) with silver is used for soil sterilization. This is a very easy, economical and efficient method.

Process

1. Fill the beds with irrigation water

2. Combine water at a rate of 3-7 ml per liter of hydrogen peroxide. (Do not mix any other chemical with it)

Apply this solution uniformly to beds, there is no need to cover the soil and only leave the surface, after 4 to 6 hours the crop can be planted (use one liter of water per one meter per square area).

Bed preparation

Raised beds to be used for the cultivation of Dutch roses because these beds provide better drainage and aeration for the root of Dutch rose.

The dimensions of beds should be as follows:

» The width of bed at the top: 90 cm

» Bed height: 45 cm

» The pathway between beds: 45 cm

Gravel sand is applied to the bottom for improved drainage. The soil is meant to be porous. Organic manure is applied to the bed as it contains plant nutrients and can enhance soil fertility. For improved root establishment FYM & DAP or SSP (10 to 15 gm / sq.meter) is ideally combined on the upper layer of the bed.

Planting

In India, for the most part, two forms of planting material are used, one has budded plants, and the other is top grafted plants. Before planting, it is necessary to pick the right dutch rose varieties and the right color combination. This enhances the increased business activities of the Dutch rose. Available in dutch rose red, pink, yellow, white, orange and bicolor, The rose flower has a strong demand in the local & export sector, and on Valentine's Day it is on the peak. Dutch rose grower earns 20 to 25 percent of sales revenue from Valentine sales.

Table 15.23: Dutch rose growers select this type of color combination in the Greenhouse

Red	50-60%
Yellow	15%
Pink	15%
Orange	10%
White	5%
Bicolour	5%

Table 15.24: The most popular Dutch rose varieties in India.

Varieties	Colour
Top secret/ Taj Mahal	Red
Gold Strike	Yellow
Corvette, Tropical Amazon	Orange
Noblesse, Sweet Avalanche	Light Pink
Bon Heur, Hot Shot	Pink
White Avalanche	White
Peach Avalanche	Peach

For transplanting selected Dutch Rose plants are usually 5 to 6 weeks old budded plants; dutch rose plants must be healthy, free from disease with well root development. The density of the Dutch plant is 8-10 plants per square meter. This is necessary. So the width of the field relies on the variety of two rows planted on one line, the gap between the field and the plant is 18 cm, and the row – the row is 30 cm.

After planting, 80 percent moisture must be kept for 4-6 weeks to prevent plant desiccation.

Special cultural practices

Cultural activities are carried out in the commercial growing of roses; this should lead to good quality and quantity of flowers; these are as follows:

Mother shoot bending

The first flower is pinched after one month from the date of planting in after the two-three-eyed buds sprout on the main branch. Such branches are producing buds. When this stage reached plant mother shoot & these branches bend to the path. Due to this cultural activity, we can get a flower of decent quality.

Plant structure development

When the mother-shoots bend after planting, the first shoot begins to develop. This shoot is used to shape the essential and strong framework of the structure of the plant. It's just been the entire life of the plant. Each of these shoots at the bottom are sliced or bent as follows:

- » Medium-shoot: cut to other or third leaflets.

- » Stronger shoot: Cut off the fifth leaf pair.

Bending in roses

Plant leaves are important for the development of nutrients, so bending is required to retain an sufficient amount of leaves on the tree. The leaves are called trees' lungs. Poor & blind shoots are picked for bending & it is a constant method and is carried out during the plant life cycle. Extract buds from the branches before bending. It is essential because bud provides shelter for thrips & botrytis of sprout buds to remove growing buds from the edges of the leaves. The following bundles of bedding are permitted to expand, and are later damaged, re-bundled if not required for growth. Bending is achieved with the first or second set of leaves.

Defoliation

The replacement of leaves is referred to as defoliation. This is primarily performed to stimulate other plant organisms to bloom or to reduce the lack of transpiration through times of stress. Defoliation can be achieved by scraping the leaves manually or by withdrawing water. Upon pruning, the shoots become defoliated.

Shoot thinning

Unproductive shoots and water suckers are sometimes discarded for high quality bulbs.

Pruning

This is suggested once a year during the second or third week of October.

Pinching

The displacement of the terminal rising component along with the stem component is called pinching. This allows to achieve high quality flowers and buds and prevents energy wastage in the production of auxiliary buds if achieved at the right level and at the right time. It's heading to apical domination.

Bud capping

The bud caps are mounted on the flower bud until they are about the size of the pea. It allows to improve the size and shape of the bud to satisfy consumer demand for the specified quality.

Disbudding

To grow high-quality flower and duration of steam side bud on this steam is separated by the name of Disbudding. Disbudding will not be done too early or too late. If done too early, it may be harmful to the leaves and if done too late for the upper leaf axil. The best time to disk when the bud hits the pea-size and reveals a slight color.

Irrigation

Water quality should be as follows:

>   » PH: Between 6.5 – 7.0

>   » EC: Between 0.5 – 1 ms / cm.

Drip irrigation specification

1. Using Pressure Compensating Dripers to maintain accuracy in the distribution of water.

2. Using 2 laterals on each of the beds.

3. Driper discharge capacity must be 1.2-4 LPH.

» After planting, irrigate the plants utilizing the micro-sprinkler method over a total of four weeks to further grow the plant uniformly. Between four weeks, this progressively switches to drip irrigation.

» The water needs of the Dutch Rose plant may be roughly 800 ml-1000 ml per plant per day.

» The dry summer sprinkler device should be used to preserve maximum humidity.

» Observes soil before irrigation and regulates soil moisture.

» After determining how much water is needed for irrigation. It's different with the season, but the frequency is the same.

» Always use drip irrigation until 12 p.m.

» As a thumb rule, the soil should be slightly moist, but it should never have excessive water. For calculate the amount of water needed for roses, using the tensiometer, pan transcript or green finger process.

Fertilization

After planting, N: P: K 20:20:20 @ 2.5 g / lit every two days for the first three months. Irrigate and fertilize regularly in limited amounts for optimal performance and also take caution to satisfy the needs of the crop. Applied Micronutrients per four days or monthly if the deficit signs of the crop raise the quantity of micronutrient. If necessary, allow a soil study every 2-3 months to establish a precise nutrient schedule. As a layman, every time you enter the greenhouse, the plants should look very healthy and shiny.

Table 15.25: Fertigation Schedule for Rose under Protected Structures

Crop Duration	Distribution pattern /ratio of fertilizers		
	N	P	K
Vegetative stage (September-October)	80	50	60
Flowering and harvesting flush (November-March)	100	60	80
Flowering and harvesting normal (April-August)	80	50	80

Pest and Disease Management

Pest management

Aphids

Aphids are really common pest. Aphids are soft-bodied insects that may be red , green, yellow, or black. They feed on very young, succulent shoots, which induce distortion. Also, aphids are kept in balance by natural predators. Alternative prevention methods include the usage of insecticidal soaps, strong water streams, or insecticides.

Japanese Beetles

These hard-shelled, metal-green, black, and gold insects can cause extensive damage to roses by their sheer numbers and voracious appetite. They like roses and flower buds, but they also target foliage. Japanese beetles are difficult to control because they are powerful fliers and continuously reinvade the area. Home gardeners often find that Sevin has the best control, but it's only up to topical. This ensures the reapplication will be carried out on a daily basis to protect the trees and flowers. Watch out for Japanese beetle traps. Traps are almost too effective and will attract a large number of beetles to the area, making the problem worse. If included, they should be put in areas away from the rose garden. Hand picking is often a recommended test for limited numbers of beetles.

Leaf Cutter Bees

It is rare to see the insects at work, but they make their existence clear by the perfectly circular holes made near to the edges of the leaves. Such leaf fragments are used to create egg partitions within the burrows. The harm they inflict is strictly cosmetic but does not warrant control.

Spider Mites

Mites are very small members of spiders. Those who can be red , black, or brown. Mites pierce the underneath of the rose leaves to drink the sap, allowing the leaves to turn gray or bronze. A fine web is a warning of a serious infestation. The mites multiply quickly, resulting in large colonies in a short period. Mites flourish in crowded, barren gardens. A high-pressure washing machine with water from a garden hose targeted to the underside of the leaves will manage mites every 2-3 days. This is going to interrupt their life cycle. Miticides, such as dicofol, assist in extreme infestations. Insecticidal soaps are also effective in the control of mites.

Thrips

Thrips are very tiny, brown insects that normally live and eat within the flowers. A deformed flower with flecked or scratched petals is usually a sign of a thrip problem. The rasping sections of the thrips trigger this damage as the petal surface is scratched to eat. Thrips are particularly attracted to yellow or light-colored roses. Any control can be achieved by using products such as orthene, malathion or insecticide soap, but even these still yield poor results. These begin to get worse at the end of June, July and August as temperatures are hot.

Rose Midge

The rose midge is a tiny fly that lay eggs in buds and shoots of roses. Larvae that develop start feeding and induce bent, misshapen or blistered flower buds and withering of stem tips. They turn black eventually. Regulation consists of pruning buds and adding insecticide if the issue continues. Midge failure usually occurs in July. Since the larvae fall to the soil for pupate, successful control is to position the weed barrier tissues under the plants to trap the larvae and keep them from reaching the soil for pupate.

Sawfly (Rose Slug)

The famous rose slug causes skeletonizing or window pane damage to rose leaves in spring and early summer. The larvae behave like caterpillars, but they are probably most closely linked to bees and wasps. Local pink slugs are green with a light brown head and sometimes have hair-like bristles. While they appear like caterpillars, BT items are not successful since they are not larvae of moths or butterflies. Regulation can involve hand-picking and the use of vegetable oils or insecticide soaps.

Diseases Management

Blackspot

This fungal disease can cause almost total defoliation of the bushes by early fall, resulting in a damaged bush on which the cane die-back and cankers become serious. Blackspot is known as circular black spots that occur on the upper surface of the leaves, beginning at the bottom of the plant and heading upwards. Infected leaves turn yellow and collapse prematurely. Each leaf spot is characterized by its fringed edge and black color from the others. Cane infections are identified as reddish-purple spots. Splashing water spreads black spots. Infection happens when leaves have been soaked for many hours, rendering it particularly dangerous during rainy seasons. Many roses are less susceptible than others, so the collection of cultivars is significant. The fungus overwintering in the fallen leaves and the stem cankers. Raking and extracting these leaves, as well as pruning the affected canes by spring until the buds swell, can help to provide some control. Do not wet the leaves while watering and place plants where there is sufficient air circulation. Fungicide spray programs must be started as soon as new leaves appear in the spring.

Management

Fungicides generally recommended for blackspot control include:

- Mancozeb
- Funginex
- Phyton 27
- Daconil 2787 or fungicides containing Daconil
- Orthenex

Powdery Mildew

Powdery mildew is a fungal disease that affects young leaves, causing them to curl and twist and produce purple coloring. As the disease progresses, the leaves are covered with white powdery fuzz. Whereas black spot is typically the most extreme in the lower part of the plant, mildew affects the upper part of the plant. Mature leaves are less likely to be impacted by this. Mildew passes through the wind and grows quickly during mild, dry days, accompanied by cold, humid nights. Infections with mildew are actually discouraged by the presence of water on the leaf. Nevertheless, holding plants wet all night to prevent mildew produces an environment that causes certain diseases to develop. Infection may be reduced by properly sanitized and fungicide spray systems. Cover both decaying and diseased canes to reduce the eventual contamination of the fungus. Because new growth is particularly susceptible, a thorough coverage of new growth with fungicide is important. Plant roses in areas where they receive good air circulation and where the foliage can dry out quickly in the early morning to prevent many kinds of diseases.

Management

Frequently used fungicides for control of powdery mildew include:

» Captan

» Funginex

» Phyton 27

» Daconil 2787

» Orthenex

Stem Cankers

There are several fungi that inflict cankers on roses. The different fungi can cause different-looking cankers, but they usually produce brown, oval-shaped, sunken, or shriveled areas anywhere on the cane. The cane dies, and the leaves wilt away from that point. Often tiny black specks can be seen on the cane surface inside the canker's boundaries. These are spore-forming fungal systems. Cankers will be pruned back every year. Cut the tissue well below the affected tissue. Protect the plant from cold or freezing injury by providing adequate winter cover. Don't cover the flowers too early in the fall. If roses are mulched until the soil freezes, moisture may be stored around the canes, which can maximize the damage caused by canker disease. Keep plants healthy with proper disease control and cultivation. Canker is a stress disease. If plants continue to grow actively, there is a better chance of avoiding cankers. There are no reliable chemical inhibitors for canker disease.

Botrytis Blight

Botrytis blight is a fungal disease that usually affects dying tissues. It is commonly found in older flowers and other parts of plants. This can also target sensitive tissues in other circumstances. Botrytis prefers hot, wet environments, sometimes allowing the disease to invade whole flowers and to develop a gray, fuzzy mold. This disease is often referred to as gray mold. Good garden drainage and the elimination of spent flowers also result in effective control of this disease. If this is not sufficient to provide adequate control, a preventive spray program may be necessary.

Management

Materials used in limiting botrytis are:

- » Mancozeb

- » Captan

- » Daconil 278

As with all spray materials, follow label directions carefully for mixing and applying.

Mosaic

The mosaic of the rose is caused by a virus. Bright yellow patterns of wavy lines can appear on the leaves of some varieties. Other varieties may not have yellow lines, but may be stunted and weak due to virus infection. Virus-infected plants can not be treated. Plant virus resistant roses, if possible. Try to control insects, particularly aphids, as they help spread the virus. When you are pruning virus-infected plants, do not prune healthy plants until you first clean the pruners. It can be achieved by dipping the blades in a 10% chlorine bleach solution and water for 60 seconds. The concentration of 25% decreases the time required to approximately 10 seconds. All infected plants should be removed and destroyed in order to reduce the spread of the virus to other plants.

Crown Gall

Crown gall is a bacterial disease that can live in the soil for 15-20 years. This induces irregularly shaped, rough, dark-colored masses (galls) to develop on stems near to the soil surface. These galls may appear as small swellings or may be several inches in diameter. Severely contaminated plants are stunted and do not mature properly. There is no successful regulation of the crown brace. Severely infected plants should be digged and removed, and roses should not be cultivated in the area for at least 5 years. Consider purchasing plants with excessive swelling or gall on lower stems or crowns. However, do not confuse the crown gall with the normal swelling that you see as a result of the budding process. Secure plants from damage to stems during cultivation. Retain vigor through fertilization and irrigation. Crown gall is not unique to roses and may influence apples, raspberries, honeysuckles, euonyms and many vegetables. For this cause, roses should not be planted where plants prone to crown gall have been eliminated due to disease. Galltrol-A, a non-pathogenic bacteria, has been used to avoid crown gall. This is also used as a dip on cane root roses before planting.

Rose Rosette

Rose rosette is becoming more widespread and may inflict serious damage. This pathogen (not yet positively identified) is spread by the eriophide mite. Symptoms include accelerated growth of shoots, the development of "witches' broom," the creation of small tufts of deformed reddish leaves and extreme thorniness. Plants are declining over time. Because affected plants cannot be healed, it is best to dig the affected plant out and destroy it. Controlling a mite has been classified as an alternative, but attempts to regulate it have proven to be inconclusive. It is very difficult to apply sprays in a timely and satisfactory manner.

Harvesting

Harvesting starts after 12-15 weeks of planting. The maximum yield is 230 flowers per square meter. (8-9 plants) a year. The cut flower has a total vase life of 10-12 days depending on the size. The best period for Harvest dutch flower is in the morning or late in the evening or during the day when the temperature is low. Bud caps 7 to 8 days before harvesting result in a 10 to 12 percent change in bud size to create a broad size head. The lower end of the flower cut stems are put in clear plastic buckets containing hot water or 500 ppm citric acid solution.

Postharvest handling

Roses cut stems must be put in a bucket of water within the polyhouse directly after harvesting and transported to a cold storage point (2-4 ° C). The length of time depends on the variety and quality of the roses. The rose flowers are ranked according to the length of the stem and the consistency of the flower bud. Graded flowers are clustered together in groups of 10, 12, 20 or 25. A collection of 20 stems is typically favored. Packing with polypropylene (24 micron) is a highly effective storage technique for roses to maintain post-storage quality and vase life.

Production Technology of Betelvine

Scientific name : *Piper betel* L.

Common name : Pan

Origin : South and South East Asia.

Introduction

The botanical name of the betel is Piper betel. It's known as 'pan' in India. Betel is a perennial, dioecious, evergreen climber that grows in the tropics and subtropics because of its leaves, which are used as a chewing stimulant. Betel is grown as an important cash crop in India. It is a spreading vine, easily rooted where the trailing stems touch the ground. The betel plant is an evergreen, annual creeper with shiny, heart-shaped leaves and white catkin. The leaves are alternate, complete, 5 to 10 cm long and 3 to 6 cm wide. The flowers are thin, shaped on pendulous spikes 4 to 8 cm long at the nodes of the vine, with spikes growing up to 7 to 15 cm as the fruit matures. Betelvine leaves and stem have a pungent aromatic flavors. Betel leaves chewing is considered to be a source of diatery calcium. Betel oil has a variety of medical applications.

Soil and climate

Tropical climate, good rainfall and a shady place are ideally adapted to its vigorous development. Soil Requirements Soil with good organic matter and drainage system is best suited for the growth of betel vines. However, it can be produced on various soil types, such as heavy clay loam, laterite and sandy loam soils.

Soil preparation

In order to protect betelvines from plant-born diseases, land preparation was considered to be the most essential function. Betelvine crop needs adequate management of moisture and humidity. The selected land had to be ploughed and harrowed twice to remove all pesticides and insecticides. Land should be made smooth to promote the growth of the betelvine plant. Approximately 15 to 25 tons of farm yard manure was used per acre. Care will be taken to match the manures with the dirt. Sheep penning was often used to increase the ability of the soil to produce more per acre. The soil should be well prepared by 4–5 ploughings and the soil should be raised by 5–10 cm from the adjacent areas, providing proper gradient on both sides for rapid drainage. Field beds of the appropriate size (15 cm high and 30 cm wide) are then prepared. Until cuttings are planted, the soil should be completely sterilized.

Soil sterilization

During hot summer months (March – May) when soil temperature rises sufficiently, the soil is covered by polyethylene sheets to destroy the inoculum of soil pathogens. The use of Carbofuran @ 1.5kg / ha or neem cake (0.5 ton / ha) + Carbofuran (0.75kg / ha) for new plantations is also recommended to minimize the initial soil nematode population. Carbofuran should not, however, be recommended at any stage in established gardens, as a time gap of 65–70 days as a safe waiting period is required between the application and harvesting of leaves.

Varieties

Depending on the shape, size, brittleness and taste of the leaf blade, betelvine is categorized into pungent and non-pungent varieties.

Table 15.26: Selection of variety under different states

State	Important variety
Andhra Pradesh	Karapaku, Chennor, Tellaku, Bangla, Kalli Patti
Assam	Assam Patti, Awani pan, Bangla, Khasi Pan
Bihar	Desi Pan, Calcutta, Paton, Maghai, Bangla
Gujarat	Bangla, Kapuri, Kari, Kalakatti, Mithi
Karnataka	Kariyale, Mysoreale, Ambadiale
Kerala	Nadan, Kalkodi, Puthukodi
Madhya Pradesh	Desi Bangla, Calcutta, Deswari
Maharashtra	Kallipatti, Kapoori, Bangla (Ramtek)
Orissa	Godi Bangla, Nova Cuttak, Sanchi, Birkoli
Tamil Nadu	Pachai Kodi, Vellaikodi
Uttar Pradesh	Deswari, Kapoori, Maghai, Bangla
West Bengal	Bangla, Sanchi, Mitha, Kali Bangla, Simurali Bangla.

Propagation

Stem cuttings of 3–5 nodes are used for propagation and cultivated in such a way that 2–3 nodes are buried in the ground surface. A single node with a mother leaf is also planted. Cuts of the apical and middle portions of the vine are used for planting.

Cultivation practices

In India, two forms of cultivation are practiced: an open method of cultivation using support plants and a closed system of cultivation using artificial rectangular constructs called barejas.

Construction of Bareja (rectangular structures) for artificial support and shade

Barejas are typically produced on slightly sloping land, near a source of irrigation at a higher level than the neighboring field. There must be a slope in both directions for the accelerated drainage of excess water. Barejas are nothing but rectangular frameworks made up of bamboo or jute sticks, typically 2–2.5 meters in height. These rectangular structures are covered with thatching by means of coconut leaves or straw or other materials.

Raising of support plants for natural support and shade

Sesbania grandiflora plants, *S. Sesban, Erythrina variegata* and *Moringa oleifera* have been grown to provide shelter and protection. These are sown in rows of 45–60 cm at least 45 days until the cuttings of betelvine are cultivated.

Planting of Betelvine cuttings

40,000–75,000 cuttings per hectare are required in an open method of cultivation, while 1,000–1,20,000 cuttings per hectare are appropriate for bareja (closed) cultivation. The onset of the monsoon is the ideal time for planting under a closed cultivation system. October is the perfect time to plant in the free method of cultivation. However, the planting season in India differs from state to state.

Table 15.27: Planting season under different states

State	Planting season
Andhra Pradesh	September October
Assam	April May and August September
Bihar	June July, September and May June
Karnataka	July August
Kerala	May June and September November

Madhya Pradesh	January March and September November
Maharashtra	July August and October November
Nagaland	April May
Orissa	May June and September November
Tamil Nadu	March May and August October
Tripura	March April and September October
Uttar Pradesh	October November
West Bengal	June July and September October

Spacing

The planting process is performed in rows. In different states, different spacing is practiced.

Table 15.28: Spacing of betelvine in different states

State	Spacing (cm)
West Bengal	50-70 x 10-20
Uttar Pradesh	100 x 10-15
Maharashtra	80 x 20
Bihar	80-100 x 10-20
Madhya Pradesh	50-60 x 15

Training and pruning

One month after planting, young shoots begin to appear and these young shoots are trained along the support and bound with them using banana fiber or jute fiber once every 15 to 20 days.

Fertilizer schedule

Oil seed cakes such as castor cake, linseed cake, sesame cake or neem cake are used as manure @ 15 Q / ha. The cake has been immersed in water for the first time in a big earthen pot for a week. This slurry is then added at regular intervals. Oilcakes in powder form are often used throughout the rainy season. Nitrogen @ 200kg / ha / year should be used as a plant manure or oil cake. A dosage of 100 kg of P_2O_5 and K_2O / ha / year is also prescribed. Fertilizers can be given in 4–6 single doses at 2–3 months intervals.

Irrigation

Betelvine provided regular but light irrigation supplies kept the land continuously moist. Farmers who had an enough supply of water during the year tended to cultivate betelvine crops. Surface irrigation was practiced in most betelvine gardens. Some gardens were on the irrigation drip system. Betelvine needed more water during the summer season, although it wanted fewer water during the rainy season. A proper surface drainage system was needed during the rainy season. Any farmers now face lack of water for a few days. Some farmers have established ponds in their farms to address this issue. Poly thin paper was spread out in the ponds. During the rainy season, so much water was collected in the farm pond and used throughout the summer season due to the scarcity of water. Excessive irrigation and stagnation of water must be prevented. Because betelvine needs high soil moisture, regular light irrigation is important depending on the season. Irrigation should be focused on the need for proper drainage throughout the rainy season.

Pest and Disease Management

Pest management

Betelvine bug (*Disphinctus poiitus*)

When sufficient measures are not taken, betelvine causes serious damage. Spray with dichlororas (Nuvan) 500 ml. in 500 liters, it will protect against disease.

» Removal of weeds and alternative host plants such as hibiscus, bhendi, custard apple, guava etc. in and surrounding vineyards during the year.

» Deep ploughing in summer or raking in vineyards tends to kill the nymphal stages and reduce the occurrence.

» Detrash the crop on DAP 150 and 210.

» Exotic predator release, *Cryptolaemus montrouzieri* @ 5 per infested plant as per literature.

Red Spider Mites

Red spider mite was the major insect that triggered the red browning of the views and valets that extended to the blade of the vine and ruined the whole leaf. Infected leaves are unsuitable for chewing. Cultivators together malathion 50 EC @ 500 ml or wettable sulphur 80 wp 1000 kg in 500 liters of water.

Scale insects

» Initiate prevention steps during the early phases of insect infestation.

» Pick the size of free seed vines.

» Common standards are as followed for others.

» In nurseries spraying neem oil 0.3 per cent or fish oil rosin 3.0 per cent is also successful in managing pest infestation.

Nematodes

Before planting, farmers should applied neemcake at the rate of two tonnes and carbofuran at the rate of 1.5 kg per hectare. This kept the nematodes infraction under check.

Diseases Management

Betelvine crop was attracted by the amount of diseases that caused significant crop losses. It was safer to cultivate this crop under the protection of plants.

a) Foot and Leaf Rot (*Phytophthora arasitica*) or collar rot (*Fusarial* wilt)

» The disease of foot and leaf rot emerged shortly after the beginning of the monsoon. A strong irrigation program has helped reduce the disease. Application by B.M. (1 per cent) during the rainy season as the soil drenches and sprays 0.5 per cent throughout the 20-30-day period, this scheme helped to regulate the system.

» The removal and destruction of dead vines along with the root system from the garden is essential as this reduces the formation of inoculum (fungal population).

» Planting content must be obtained from disease-free gardens and nurseries ideally grown in fumigated or solarized dirt.

» Adequate drainage to minimize water stagnation should be given.

» Injury against the root structure owing to traditional activities such as digging should be stopped.

» Freshly developing runner shoots should not be permitted to track on the ground. They will either be tied up to the standard or they must be pruned off.

» Around the start of the monsoon, the leaves of the support trees must be pruned in order to prevent the build-up of humidity and greater penetration of the sunshine.

» Reduced humidity and the presence of sunlight reduces the intensity of leaf infection.

» Copper oxychloride 50 percent WP @ 1 Kg in 300-400 liters of water / acre.

» Eradication of the affected vineyard.

» Add the phytosanitary process.

Leaf Spot: Anthracnose (*Collecfotrichum capsid*)

Brown or black spot with visible concentric circles formed on the leaves. The disease was controlled by spraying 0.5 per cent or 0.3 per cent of copper oxychloride in the Bordeaux mixture at a 20-day interval.

Bacterial Leaf Spot

Betelvine was directly infected by a bacterial leaf mark, which was a severe disease. They emerged on the lower surface and the yellow halo on the matching upper surface of the vine. The disease was handled by administering the Bordeaux combination at 0.5 per cent or 0.3 per cent with copper oxychloride at 20-day intervals.

» Collection and disposal of contaminated plant pieces minimizes the transmission of the disease.

» Improve the movement of air in the vineyard.

» Remove disease cane from the vineyard during regular pruning during the dormant season.

» Follow up the hand pruning.

Sclerotlal wilt (*Sclevotium rolfsii*)

In this scenario, the stems may become dark. The leaves dropped and gradually the plants were dead, and then they died. The disease was regulated by the usage of green manure, organic and inorganic manures. Spraying at 0.5 per cent of the Bordeaux mixture or 0.3 per cent of copper Oxychloride at 25 liters at 20-day interval provides stronger performance.

Powdery mildew

» Appropriate cultural activities that minimize humidity inside the vineyard, enable for good air movement through the canopy and provide sufficient light access to all the leaves and clusters that assist in the control of powdery mildew.

» Using the under vine irrigation device and control it carefully, the excess can benefit the disease.

Harvesting and Postharvest Management

Once the vines exceed a certain height, the leaves are collected from the lower part of the plant. Harvesting takes place during March – April in Uttar Pradesh, Madhya Pradesh and Bihar, during May – June in Andhra Pradesh, and either January – February or April – May in Tamil Nadu. Mature leaves are plucked along with a petiole. They're picked up by hand. In Karnataka and Tamil Nadu, the leaves are plucked by side shoots. Comparatively tender leaves are preferred on the market in southern India. They are washed thoroughly after plucking and made into bundles according to the prevailing traditions of the region. On average, 60–80 lakh leaves are collected annually from one hectare of plantation.

Harvested leaves are washed, cleaned and rated according to size and quality. They are packed after removing a portion of the petiole and rejecting the damaged leaves. The selected leaves are sorted into different grades according to size, color, texture and maturity. After all this, they're grouped in packaging numbers. Typically bamboo baskets are used for wrapping and, in certain cases, straw, fresh or dry banana leaves, wet cloth, etc. are used for inner lining. Typically betel leaves are used as young, unprocessed chewing. For certain places, though, leaves are subject to processes such as bleaching or curing. There is a good demand for these leaves, which brings higher prices to the markets. Bleaching is achieved by sequential heat treatments at 60-70°C for 6–8hrs.

Chapter - 16

Drying Methods and Applications for Hi-tech Horticulture

Introduction

There would be no time difference between crops in an effectively controlled greenhouse CEA. However, for certain other management purposes, if crops are not produced within a specified time span, the greenhouse may be used as a solar dryer. A significant volume of 15 to 30 per cent of the received solar radiation is mirrored back from the top of the greenhouse and the rest is transferred to the interior. The bulk of this emitted radiation is consumed by trees, vegetation and other interior objects, the remainder being mirrored. The usage of greenhouses for drying purposes is of modern history. They have been successful in promoting the year-round use of the greenhouse facility, thus reducing operating costs per unit output. In general, the product is spread as thin layers in trays going to cover the greenhouse area. The trays may be constructed of sheet metal and wire mesh. Trays can be horizontally placed on current rising benches or frames. For better functioning, adequate ventilation should be provided either by forced or natural ventilation, by extracting moisture from the substance and by regulating the temperature of the air within the greenhouse. Natural airflow may be improved with the usage of a black LDPE chimney attached to the greenhouse.

Greenhouse dryer

The concept of a greenhouse dryer is to merge the role of a solar collector with a greenhouse device. The roof and wall of this solar dryer may be constructed of translucent materials such as glass, fiberglass, UV-stabilized plastic or polycarbonate sheets. Transparent fabrics are fixed to the framework of a steel frame or to pillars with bolts and nuts and locking to avoid wet air or rainwater from escaping into the chamber rather than those released from the opening of the inlet. Black surfaces should be given inside the framework to improve the absorption of solar radiation. Inlet and exhaust ventilators are positioned in the correct location inside the system to ensure equal delivery of the drying air. If properly built, greenhouse dryers can have a greater degree of influence over the drying cycle than cabinet dryers and are more suited for large-scale drying.

Ekechukwu has built a natural convection greenhouse dryer consisting of two parallel lines of drying platforms (along the long side) of galvanized iron wire mesh placed on wooden beams. The fixed inclined glass roof over the platform allowed the product to be exposed to solar radiation. The dryer, positioned lengthwise in the north-south line, had black coated inner walls for better absorption of solar radiation. A ridge cap consisting of folded zinc sheet over the roof creates an air outlet. The air inlet was controlled by shutters on the outer sides of the platforms.

A simpler version of a traditional greenhouse-type natural circulation solar dryer consists of a translucent semi-cylindrical drying chamber with a cylindrical chimney attached, growing vertically from one end, while the other end is fitted with an air inlet door and entry to the drying chamber. The chimney (designed to accommodate for different heights) has an overall height of 3.0 m above the chamber and a diameter of 1.64 m. The drying chamber was a redesigned and improved variant of the commonly accessible poly tunnel style greenhouse.

The dryer works as a function of the operation of solar energy working directly on the component within the dryer. The product and the vertically ended up hanging black absorbing curtain inside the chimney soak up the solar radiation and are heated, which in turn heats the surrounding air. When the warm air rises and passes up the chimney to the outside of the dryer, new refilling air is pulled from the other end of the dryer. Apart from the apparent benefits of passive solar-energy dryers over active forms (for rural farm applications in developed countries), the benefits of the natural ventilation of solar-energy ventilated greenhouse dryers over other passive solar-energy dryers include its low cost and flexibility of both on-site installation and service. The biggest downside is its vulnerability to disruption at strong wind levels.

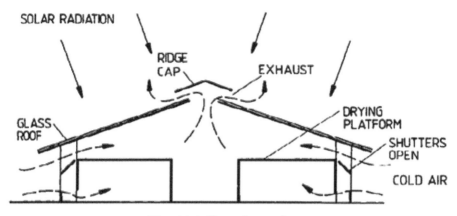

Fig. 16.1: Greenhouse dryer

Types of solar dryers

1. **Tent dryer:** Tent solar dryers are inexpensive and simple to construct and consist of a frame of wooden poles lined with plastic sheeting. Black acrylic is to be included on the wall faced away from the light. The substance to be dried is put on the soil above the field. It takes the same amount of time for the drying of products as for the drying of open air. The primary function of dryers is to shield them from dust, soil, heat, wind or pests and are typically used for berries, seafood, coffee or other items for which the waste is otherwise large. Tent dryers may also be disconnected and placed while not in operation. These have the downside of being quickly destroyed by strong winds.

2. **Box dryer:** The box-type solar dryer has been commonly used for small-scale food drying. It comprises of a wooden box with such a transparent lid attached to it. The interior layer is painted black and the device is protected by a mesh tray over the dryer board. Air flows via the holes in the front chamber and leaves from the vent holes at the top of the back wall.

3. **Solar cabinet dryer:** The cabinet is a large wooden or metal box and the commodity is stored in trays or shelves inside a drying cabinet. If the chamber is translucent, the dryer shall be referred to as an integral form or direct solar dryer. Unless the chamber is opaque, the dryer is referred to as a dispersed form or indirect solar dryer. Mixed-mode dryers incorporate the characteristics of the integral (direct) form and the dispersed (indirect)

model of solar dryers. The combined effect of the solar radiation incident directly on the substance to be dried and the hot air generates the requisite heat for the drying phase. For most situations, the air is heated during its passage via a low-pressure solar collector and moves through the air ducts into the drying chamber and then through the drying trays containing the crops. The hot air is then expelled via the air vents or the chimney at the top of the room. It should be well insulated to mitigate heat losses and rendered robust (within economically justifiable limits). Construction of metal sheets or water resistant coatings, e.g. paint or resin, is preferred.

Heated air passes through the tray stack until the whole package is hot. When the hot air flows into the bottom tray, the tray must dry first. The last drying tray is the one at the top of the container.

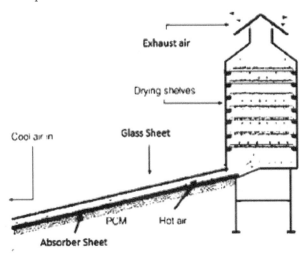

Fig. 16.2: Solar cabinet dryer

PHT equipment's design and operation

1. All devices should have those qualities:

2. It is expected to perform the role for which it is built.

3. Its work should be completed cleanly, and the computer should be washed quickly.

4. The machine should be built as simply as practicable and of such durable construction that little modifications are required.

5. It is supposed to be economical in service.

Food plant equipment design

1. All machine parts must be built for easy disassembly and reassembly – some simply by removing and replacing a nut or a wing screw by hand. It is often possible to build these pieces of lightweight content in such a way that they can be quickly washed.

2. All food touch surfaces should be neutral, shiny, non-porous and non-absorbent; they must survive the application of industrial chemicals, sanitizers and pesticides; they must be quickly washed and readily available for examination.

3. Open seams for boiling kettles, blenders, blenders, storage vats and filled devices must be withdrawn.

4. The surfaces of food-contact equipment must be smooth and continuous; rough spots and cracks must be avoided.

5. All junctions, in specific pipes and ducts, must be angled or squared. Cooking kettles, storage containers, storing vats and related devices will have large curves at the corners of the bottom and side walls instead of straight edges.

6. All equipment pieces in touch with food must be accessible for hand brush cleaning and/or examination.

7. Dead-end places will be removed in all machines.

8. Metals such as arsenic, antimony and cadmium will not be used in industrial machinery. Copper or copper alloys are not suggested.

9. Stuffing boxes or glands in which food may accumulate and decompose should not be used.

10. Pipe fittings must have a sanitary thread and threaded parts must be made available for cleaning.

11. Sanitary-style openings, such as the form of screw, can be included.

12. Runoff valves will be mounted as near as possible to mixers, boilers, vats and reservoirs.

13. Coupling nuts on pipes and valves must have adequate clearance and must be quickly extracted.

14. Food items should be shielded from lubricants and condensates, because moisture condensing on ceilings can pick up dirt and peeled paint, then drop into open boilers or carry vats.

15. The mixing blades will either be welded to the drive shaft or both should be in one piece. Shaft and blades will be separated from the mixer at a level above the liquid line.

16. Device components in touch with food can be made of non-corrosive materials.

17. Equipment such as kettles, certain mixers, and vats and storage bins should have sectional covers that are free from the seams, hinges, cracks, and heads in which the dirt may be collected.

18. Drive shafts must be covered in such a manner that the lubricating oil does not make its way through food.

19. The horizontal sections of machinery or supports should be at least 15 cm (6 in) off the surface. Tubular supports are favoured, although if square tubes are mounted horizontally, they can be rotated at 45 ° to avoid flat surfaces. Equipment covering broad floor areas would have a height of 46 cm (18 in) or more in order to enable washing.

Cleaning and grading, methods of grading, equipment for grading

Both nutritional raw materials are washed prior to manufacturing. Clearly, the goal is to eliminate pollutants that vary from harmless to hazardous. It is necessary to remember that the elimination of pollutants is crucial both for the safety of process machinery and for the final user. For starters, rocks, stones or metallic particles must be separated from wheat before milling to prevent damaging the machinery.

The major contaminants are the following:

» Unwanted sections of the herb, such as seeds, twigs, husks.

» Sun, rain, stones and metallic particles from the rising area.

» Insects and their eggs.

» Animal excreta, feathers, etc.

» Pesticides and fertilizers.

» Mineral oil.

» React to microorganisms and their poisons.

» Cleaning is basically a division in which certain variations in the physical properties of pollutants and food components are used.

» A variety of cleaning methods are usable, listed as dry and wet methods, although a mixture will usually be required on any particular item.

» The option of effective cleaning procedure depends on the substance being washed, the extent and form of pollution and the degree of decontamination needed. In fact, a compromise must be found between the cost of cleaning and the consistency of the product and an appropriate specification for the actual end-use should be defined. Avoidance of product harm is a significant contributing factor, particularly in fragile materials such as soft fruit.

Dry Cleaning Methods

The primary dry cleaning strategies are focused on screens, aspiration or magnetic isolation. Dry methods are usually safer than wet methods because the effluent is easier to dispose of, but they appear to be less effective in terms of washing efficiency. Recontamination of the substance with dust is a big concern. Precautions may be required to reduce the possibility of dust explosions and fires.

Screens are simply scale separators based on perforated beds or wire mesh. Larger pollutants are extracted from tiny food products. Eg: straw from rice plants, or pods and peas twigs. This is called scalping. Rotary drums and flatbed shapes are the predominant geometries.

Abrasion, either by effect during computer service or by use of abrasive disks or brushes, may increase the performance of dry displays. Screening results in incomplete separations and is typically a basic cleaning step. Aspiration uses the variations in the aerodynamic qualities of food and contaminants. It is commonly used in the washing of cereals, but is often used in the washing of peas and beans.

Recirculatory Batch Dryer (PHTC type)

It is a constant flow of a non-mixing form of grain dryer. The dryer consists of two concentric circular cylinders constructed of 20 gauge perforated (2 mm dia) sheet steel. The two tubes measure 15 to 20 cm apart. Such two cylinders are backed by four portions of the pipes. The whole frame may be protected by an adequate foundation or can be bolted to a frame consisting of a channel segment. A appropriate size bucket elevator is used to feed and recycle the grain into the dryer. A centrifugal blower blasts hot air through the inner cavity that serves like a plenum. The hot air from the plenum travels through the grain passing downwards through friction and emerges out of the outer perforated cylinder. A torch burner is used to provide the required heat with kerosene oil as fuel. PHTC dryer styles with a keeping capacity of 1/2, 1 and 2 tonnes are usable. The 2-ton PHTC dryer holding capability established at PHTC, IIT, Kharagpur, India is shown in fig.

The grain is pushed to the top of the inner container. When falling through the annular space from the feed end to the discharge end through gravity, the grain comes into contact with a cross-flow of hot air. The waste air is discharged through the perforations of the exterior cylinder and the grain is discharged through the outlet of the hopper. The feed rate of the grain shall be regulated by the closing or opening of the gate with the outer pipe of the discharge hopper. The grain is recirculated until it is dry to the appropriate moisture amount.

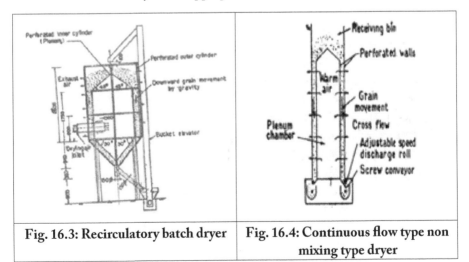

| Fig. 16.3: Recirculatory batch dryer | Fig. 16.4: Continuous flow type non mixing type dryer |

Advantages

1. Price is a fair rate.

2. The easiest feature of all flow style dryers

3. Simple to use

4. This may also be used on the field and on the rice mill.

5. Operating costs are small with the furnace fired from the husk.

Disadvantages

1. Drying is not as standardized in contrast to the mixing process.

2. The perforations of the cylinders can be clogged with the parboiled paddy after a long period of use.

Louisiana State University Dryer

It is a continuous flow-mixing form of grain dryer that is common in India and the U.S.A.

1. A rectangular drying chamber equipped with air ports and a storage bin;

2. An air blower with a pipe,

3. Grain discharging device with a bottom hopper, and

4. The air heating system.

1) Rectangular bin: Typically, the following top square portions of the bin are used for the LSU dryer configuration. (i) 1.2 m x 1.2 m, (ii) 1.5 m m x 1.5 m, (iii) 1.8 m m x 1.8 m and (iv) 2.1 m x 2.1 m. The rectangular bin can be separated into two parts, including the top holding bin and the bottom drying chamber.

2) Air distribution system: Layers of inverted V-shaped canals (called inverted Vports) are mounted in the drying room. Heated air is delivered into the falling grain bulk through these channels at several stages. There is a gap on one end of each air pipe, and the other end is covered. Alternate materials are inlet air and drain air pipes. Throughout the inlet layers, the channel openings meet the air inlet plenum area, but are sealed on the opposite wall, where, as in the outlet layers, the channel openings meet the exhaust but are sealed on the other side. Inlet and outlet ports are positioned in an offset design one below the other. Therefore, air is pushed into the falling grain as it travels from the feed end to the discharge point. Inlet ports consist of a few full-size ports and two half-size ports on both sides. Both these ports of the same scale are placed in equivalent space between them. The amount of ports that hold a dryer differs greatly based on the scale of the dryer. Every layer is offset in such a way that the top of the inverted V ports helps to separate the grain stream and flow the grains through such ports along the zigzag path. Heated air is provided by a blower in most versions.

3) Grain discharging mechanism: Three or four ribbed rollers are provided at the bottom of the drying room, and can be rotated at varying low speeds at various grain discharge levels. The grain is discharged into a hopper connected to the bottom of the drying container. Thanks to the mixture of grain and air, the discharge mechanism at the base of the dryer often controls the rate of dropping of the grain.

4) Air heating system: Air is heated by burning gaseous fuels such as natural gas, butane gas, etc., or by liquid fuels such as kerosene, furnace oil, heating oil, etc., or by solid fuels such as wood, husk, etc. Heat may be provided explicitly by the usage of a gas burner or oil burner or a furnace burner and indirectly through the use of heat exchangers. Indirect heating is often less effective than direct heating. Nonetheless, oil-fired burners or gas burners could be directly substituted by husk-fired furnace for the economy of grain drying. Heated air is added at several stages in the dryer so that it can be circulated uniformly between the inlet ports and the falling grain mass. It exits via the ports of the outlet. This form of dryer is often fitted with a special fan to blast the ambient air from the bottom cooling portion through which the dried or partly dried hot grain comes into contact with the ambient air. In general, the dryer size ranges from 2 to 12 tons of grain, although often higher capacity dryers are also installed. The power demand accordingly differs widely. The required air flow range is 60-70 m^3/min/tonne of parboiled paddy and the optimal air temperatures are 60^0C and 85^0C for raw and parboiled paddy. A variety of dryers may also be mounted.

Advantages & Disadvantages

1. It is possible to produce uniformly dried goods if the dryer is correctly built.

2. The dryer may be used for various grain forms.

3. Large spending in energy

4. Drying costs are very large if oil is used as a source.

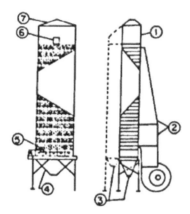

1. Garner

2. Duct

3. Dry material outlet

4. Hopper

5. Continuous flow

6. Door

7. Roof

Fig. 16.5. LSU dryer

Tray Dryer

The design of the standard batch dryer is shown in fig. Tray dryers typically work in batch mode, using racks to carry the commodity and pump air through the material. It consists of a rectangular sheet metal chamber comprising trucks supplying the bar. Growing rack holds a variety of trays filled with the material to be dried. Cold air passes into the tube into the bar. Fans are often used to blast hot air through the trays on the surface of the tube. And the baffles are used to spread the air equally over the tray deck. Some hot air is continually ventilated through the exhaust duct; new air is collected through the inlet. The racks holding the dry product shall be brought to the tray-dumping site.

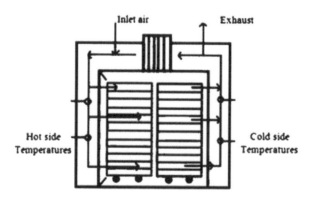

Fig. 16.6. Tray dryer

Such forms of dryers are helpful when the output volume is small. These are used to dry a broad variety of products, but have heavy labour costs for loading and unloading goods and are costly to work. We considered the most common method for the drying of useful items. The drying process in the event of these dryers is sluggish and takes multiple hours to complete the drying of one sample. For indirect ventilation, dryers may also be run under a vacuum. The trays may rest on hollow plates filled with steam or hot water, or may contain space for heating oil. Vapour from the firm may be extracted by an ejector or a vacuum pump.

Drum Dryer

For drum dryers, a solvent comprising dissolved solids or slurry comprising condensed solids forms a thin film on the outside surface of a large spinning drum. For a single drum device, the film thickness may be managed by an adjustable scraping tool. The distance between the drums may be regulated in the case of a double drum device size. A fluid, usually rain, may be sprayed across the surface for quick removal of moisture. The movement of the drum is balanced such that all the solvent is completely vaporized and the dry residue may be scraped off with the aid of a compact or customizable knife. It form of dryer is primarily used for fabrics that are too dense for a spray dryer and too small for a rotary dryer. The strong gathers on the apron in front of the knife and falls into a jar or a screw conveyor. The function of the drum dryer is constant. The drum is constantly rotated by a mechanism driven by a pinion, which derives its motion by a loop, chain or reduction mechanism. The speed of the drum can be controlled by a variable-speed drive to adapt the speed to any minor difference in the consistency of the feed. The speed of the drum is governed by the quality of the products (i.e. wet or dry) if the substance of the component becomes wet / dry such a while until the knife becomes hit, the speed will be reduced / increased. The configuration of the modules is identical to the nature of the drum filter. The knife may only be kept to the wall. This may be carried forward by rotating the changing pedals. The knife supports should be rotated around a section of the circle such that the angle of the knife edge relative to the drum surface can be selected for the best shearing effect. Throughout recent years, dual drum dryers have substituted single drum dryers in a variety of applications owing to their more reliable service, a wide range of goods and high output speeds.

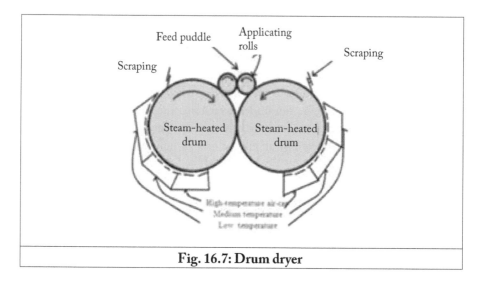

Fig. 16.7: Drum dryer

Fluidized Bed Dryer

Fluid bed dryer consists of a cylindrical or rectangular cross-section steel frame. A grid is given throughout the column on which the wet content rests. In this form of dryer, the drying gas flows through the solid bed at a pace adequate to hold the bed in a fluidized condition. Mixing and heat transfer is really easy in this form of dryer. The dryer may be worked either in batch or in continuous mode. Fluid bed dryers are ideal for granular and crystalline products. When fine particles are present, either from the feed or from the particle breakage in the fluidized surface, there could be significant solid transfer with the outlet gas and bag filters are required for the recovery of fines. The key benefit of this style of dryer is: fast and consistent heat transfer, quick drying period and strong control of drying conditions. Throughout the case of rectangular fluid-bed dryers, different fluidized compartments are created from which the solids pass sequentially from inlet to outlet. We are known as plug flow dryers; the time of life is about the same with all the objects in the compartments. Nevertheless, the drying conditions may be changed from one compartment to another, and sometimes the last compartment is fluidized with cold gas to cool the solid before discharge.

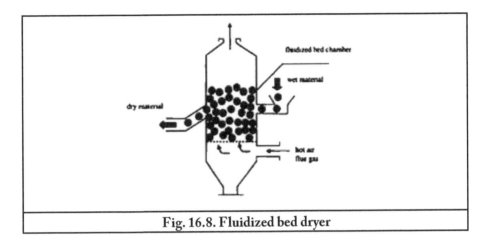

Fig. 16.8. Fluidized bed dryer

Spray Dryers

Foods are converted from slurry to dry powder in a spray dryer. A fine dispersion of pre-concentrated food is first atomized to form droplets (10-200 μm in diameter) and then sprayed into a hot air stream at 150-300 °C in a wide drying chamber. The spray drying operation is easily divided into three distinct processes: atomization, drying by contact between the droplets and the heated air, and the collection of the product by separating it from the drying air.

When the liquid food droplets travel with the warm air, the water evaporates and is taken away by the breeze. Much of the drying happens over a constant-rate cycle and is constrained by mass transfer to the droplet surface. Upon meeting the essential moisture level, the composition of the dry food particle determines the time of drying down. In this phase of the process, the diffusion of moisture inside the sample is a rate-limiting parameter.

Once the dried food particles exit the drying container, the substance is isolated from the environment by a cyclone separator. The dry commodity is then put in a sealed container with a moisture content normally below 5%. Material consistency is known to be outstanding thanks to the preservation of oil solids by evaporative cooling in the spray dryer. The small particle size of the dried solid facilitates simple reconstitution when mixed with water.

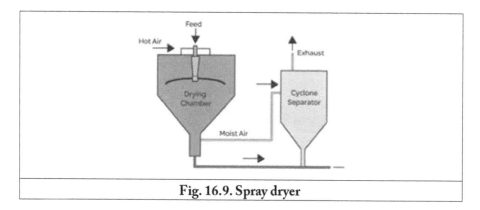

Fig. 16.9. Spray dryer

Freeze Dryers

Freeze-drying is accomplished by reducing the temperature of the surface such that much of the product's moisture is in a stable state and by decreasing the friction inside the component, the sublimation of the ice can be done. If product consistency is a significant consideration for market adoption, freeze-drying offers an alternate solution to moisture elimination.

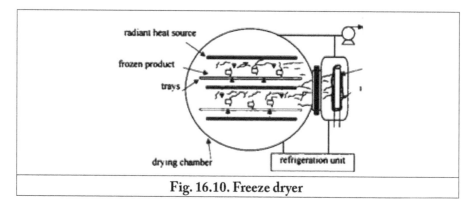

Fig. 16.10. Freeze dryer

Principle of aspiration cleaning

The idea is to feed the raw material through a carefully controlled upward air current. Denser material will fall, while lighter content may be carried away based on the strength of the terminal. Terminal velocity in this case can be specified as the velocity of the upstream air in which the particle stays stationary; and this depends on the density and predicted region of the particles (as stated in Stokes' equation). When utilizing varying air speeds, it is possible to distinguish, say, wheat from lighter chaff or denser small stones. Quite precise separations are possible, but a significant amount of energy is needed to produce air streams. Clearly, the device is constrained by the scale of the raw material units, but is especially appropriate for the cleaning of legumes and cereals.

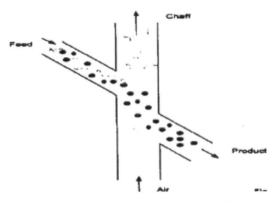

Fig. 16.11: Principle of aspiration cleaning

Magnetic cleaning

Magnetic cleaning is the processing of the given metal by means of permanent or electromagnets. Metal particles, extracted from the rising field or pic ked up during transport or preliminary activity, present a danger to both the customer and the processor, for example, cereal mills. The configuration of magnetic cleaning systems can be very variable: particulates may be moved over magnetized drums or magnetized conveyor belts, or strong magnets may be mounted above the conveyors. Electromagnets are simple to clean when shutting off electricity. Pre-sensitive processing machinery is also utilized for metal detectors as well as to secure users at the end of production lines.

Electrostatic cleaning

Electrostatic washing can be used in a small range of situations where the surface load on raw materials varies from the contaminating particles. The theory that be used to differentiate grains from other seeds with similar morphology but with specific surface loads; and it has also been defined for the cleaning with tea. The feed is transmitted on a balanced belt and the charged particles are drawn to the opposite electrode due to their surface charge.

Wet cleaning methods

» Wet methods are important if significant amounts of soil are to be removed; and they are vital if detergents are to be used. Nevertheless, they are costly, because large quantities of high-purity water are needed and the same quantity of dirty effluent is generated.

» Treatment and reuse of water can reduce costs. The usage of the counter current principle will minimize water demands and effluent volumes if it is precisely regulated.

» Sanitizing chemicals such as chlorine, citric acid and ozone are widely used in washwater, particularly in conjunction with peeling and reduction in thickness, where reducing enzymatic browning can also be a goal. Levels of 100–200 mg l–1 chlorine or citric acid can be used, but their decontamination effectiveness has been disputed and is not allowed in certain countries.

Soaking is a basic step in the washing of highly polluted products, such as root crops, for soil softening and the gradual elimination of stones and other pollutants. Metallic or concrete containers or drums are used, and can be equipped with water agitation aids, like stirrers, paddles or revolving structures for the entire drum. With fragile items such as strawberries or asparagus, or goods that collect soil within, e.g. celery, preserving air via the device can be beneficial. The use of warm water, including using detergents, enhances cleaning performance, particularly where mineral oil is a possible contaminant but adds to the expense and may affect the texture.

Spray washing

Spray washing is commonly used by other forms of agricultural raw materials. Performance depends on the quantity and temperature of the water and the exposure period. As a general rule, small amounts of high-pressure water offer the most effective removal of soil, but this is constrained by damage to the substance, particularly to more fragile items. For larger food bits, it may be possible to rotate the device in such a way that the entire surface is presented to the spray that it may be essential to rotate the device in such a way that the entire surface is presented to the spray. Drum washers and belt washers are the two most popular styles. Abrasion can add to the cleaning effect, but must again be restricted in delicate units. Many prototypes have lightweight rubber disks that softly clean the floor.

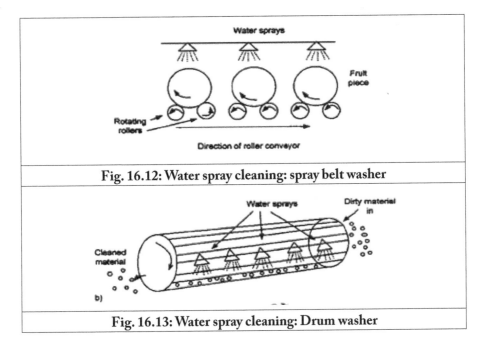

Fig. 16.12: Water spray cleaning: spray belt washer

Fig. 16.13: Water spray cleaning: Drum washer

Flotation washing

Flotation washing utilizes buoyancy variations between food units and pollutants. For eg, sound fruit is usually flat whereas dirt, stone or rotting fruit is polluted and sunk in water. As a consequence, fluming fruit in water through a set of weirs provides very productive washing of grapes, peas and beans. High water usage is a drawback and hence the recirculation of water will be integrated. Froth flotation is carried out in order to isolate the peas from the infected weed see and to maximize the surfactant results. Peas are dipped into an oil / detergent emulsion and air is expelled across the pillow. It creates a paste that washes away the contaminant content, so the cleaned peas may be washed again.

Sorting

Sorting is the division of products into groups on the basis of observable physical properties. As for washing, sorting will be used as early as possible to maintain a consistent output for eventual processing. The four primary physical properties used for food processing are scale, form, weight and colour.

Shape and size sorting

The distribution of the particle size of the substance is represented either as a mass fraction of the substance retained in each sieve or as a total percentage of the material retained. The form of such foods is critical in deciding their suitability for processing or their retail value. For eg, in the case of economic peeling, potatoes should have a standard oval or circular shape without protuberances. Sorting of shapes is performed either manually or mechanically. Scale sorting (called sieving or screening) is the division of solids into two or more fractions on the basis of variations in thickness. This is especially relevant when the product is to be heated or cooled, because the amount of heat transfer is partially dictated by the size of the individual pieces and the difference in size may induce over-processing or under-processing. In comparison, products with a standard variety are stated to be favoured by customers. Screens with set or variable apertures are used for scale sorting. The device may be fixed or, more generally, spinning or vibrating.

Fixed aperture screens

Two popular forms of fixed aperture displays are a flat screen (or a sieve) and a drum screen (a rotary screen or a reel). The multi-deck flatbed panel (Figure 5.3) has a variety of inclined or horizontal mesh displays with opening sizes varying from 20 μm to 125 mm, mounted within a vibrating frame. Food objects that are smaller than panel openings move through under gravity before they enter a computer with an opening size that holds them. The smallest objects that are commonly isolated are on the size of 50 μm.

Where friction alone is inadequate to adequately isolate the particles, a gyratory movement is used to disperse the food over the whole sieve region, and a vertical jolting motion breaks down the agglomerate and dissolves the particles that obstruct the sieve openings.

Most forms of drum displays are used to filter small-particle objects (e.g. almonds, peas or beans) that have adequate mechanical power to survive tumbling inside the device. Drum displays are nearly horizontal (5–10° inclination), wire perforated or mesh tubes. They can be concentric (one within the other), parallel (foods exit one panel and join the next) or series (a single drum built from parts of various sizes of openings). Both forms have a higher potential than flatbed screens, so blindness issues are less severe than flatbed screens. The efficiency of the drum screens decreases with their rotating speed to a crucial level. Over this, food is pulled by centrifugal force against the device and results in low separation. Similarly, there is an improvement in the power from the tilt of the panel to the vital point. Despite that, the time of residency is so limited and the goods move around without separating.

Variable-aperture screens

Variable-opening displays have either a constantly diverging opening or a stepwise rise in aperture. These styles treat food more carefully than drum displays and are also used to arrange fruit and other items that are easily harmed. Continuously variable displays use sets of diverging rollers, wires or felt-lined conveyor belts. They may be powered at varying speeds to move the food and thereby balance it, to show the smallest dimension to the opening.

Stepwise aperture changes are accomplished by changing the distance between the powered rollers and the inclined conveyor belt (see Figure on belt and roller sorter). The food rotates and the same dimension is then used as the basis for sorting (for example, the diameter around the fruit core).

Grading

This concept is sometimes used interchangeably with sorting, but specifically applies to the measurement of the general consistency of a product using a variety of attributes. Sorting (i.e. division on the basis of one feature) may then be used as part of a grading process, but not vice versa. Grading is done by operators who are qualified to evaluate a variety of factors at the same time. For starters, eggs are externally examined over tungsten lamps (so-called candles) to determine up to 20 variables and eliminate those that are, for starters, fertilized or malformed, and those that have blood spots or rot. Meats, for example, are inspected by disease regulators, fat distribution, bone to meat ratio and carcass size and form. Many rated foods like cheese and tea, which are measured for texture, flavour, colour, etc. Apples are evaluated with the aid of coloured cards displaying the necessary characteristics of various grades in terms of distribution of colour over the fruit, surface blemishes and the size and shape of the fruit.

For certain situations, the classification of food is calculated by the findings of laboratory analyses (for example, wheat flour is measured for protein content, extensibility of the grain, colour, moisture content and insect presence). In general, classification is more costly than sorting because of higher prices for professional operators. Nevertheless, other characteristics that cannot be tested automatically can be evaluated at the same time, resulting in a more standardized, and high-quality product.

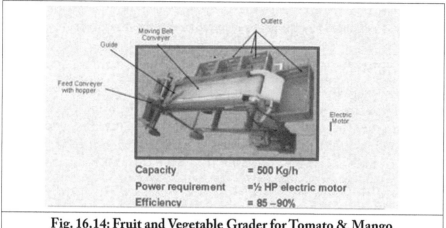

Capacity	= 500 Kg/h
Power requirement	= ½ HP electric motor
Efficiency	= 85 – 90%

Fig. 16.14: Fruit and Vegetable Grader for Tomato & Mango

Chapter - 17

HACCP, GMP, GAP and Quality standards for Hi-tech Horticulture

HACCP

HACCP stands for Hazard Analysis and Critical Control Points. Hazard is a biological, chemical or physical agent which, in the absence of control, is reasonably likely to cause illness or injury. Hazards may be destructive microorganisms or chemical and/or physical contaminants. In order to ensure safe food, the HACCP aim is to recognize hazards, establish controls and supervise these controls. HACCP is not a reactive, but rather a precautionary hazard control system. Food processors can use this to provide consumers with safer food products.

Its attempts to supply food to the U.S. Space Program in the early 1960s. This is not a null-risk system, but it is developed to reduce the risk of food contamination. In an evaluation of the performance of food regulation in the United States, the National Academy of Sciences (NAS) recommended in 1985 that certain regulatory bodies should adopt the HACCP approach and that it should be compulsory for food processors. Since then, this model has been implemented worldwide to ensure food safety.

HACCP is a precautionary food safety system, but this is not a stand-alone system. HACCP should be developed on current food safety programs like Good Manufacturing Practices (GMPs) and many others to make it work.

HACCP plan: In order to carry out a risk assessment for the creation of the HACCP plan, food processors should gain a required to work knowledge of potential risks. The HACCP framework is developed to control all highly probable food safety hazards.

These hazards are classified into three categories: biological, chemical and physical.

» **Biological hazards**

These hazards may arise from raw resources or from food processing steps that used to create the final product. Microorganisms live all across: air, dirt, fresh and salt water, skin, hair, animal skin and plants. Microorganisms are divided into different groups. Some of the major food groups involve yeasts, molds, bacteria, viruses and protozoa. Even though there are thousands of types of microorganisms, only a few pose hazards to human beings. Without adequate food, water and temperature, microorganisms will stop growing and multiplying. Some of them die and others stop working until they get the elements they need. Some methodologies of preservation, like drying or smoking, water control or food nutrients, start making these essential elements inaccessible to microorganisms.

» **Chemical hazards**

Chemical contamination can occur at any stage of food production and processing. Chemical compounds can be beneficial and used intentionally with certain foods, such as preservatives. The existence of a chemical do not always pose a risk. The quantity of the chemical could determine whether or not this is a hazard. Some may require exposure to toxic effects over long periods of time. For some of these contaminants, regulatory limits are set.

Chemical hazards can be divided into three categories:

» Naturally occurring chemicals.

» Intentionally added chemicals.

» Unintentionally or incidentally added chemicals.

» **Physical hazards**

Physical dangers include any possibly dangerous extraneous substance that is not usually found in food. If a customer absorbs foreign material or items improperly, they are liable to experience pain, illness or other negative health consequences. This is also easy to find the cause of the threat.

Guidelines for the Application of the HACCP System

Prior to the implementation of HACCP to every segment of the food chain, this industry will function in compliance with the Codex General Principles of Food Hygiene, the Codex Codes of Practice and relevant food health regulations such as FSSAI. Performance management is necessary to implement an effective HACCP system.

Throughout danger detection, assessment and associated operations in the design and installation of HACCP systems, attention must be provided to the effect of raw materials, additives, food processing methods, the role of hazard management manufacturing processes, the possible end-use of the drug, customer categories of concern and epidemiological data relevant to food health.

Application

The implementation of the HACCP principles comprises of the following functions as the Conceptual Order for the Implementation of the HACCP Principles.

The HACCP framework is based on the following seven principles:

- » **Principle 1**: Practice risk mitigation.
- » **Principle 2**: Place Important Control Points (CCPs).
- » **Principle 3**: Define vital boundaries.
- » **Principle 4**: Establish a CCP control framework.
- » **Principle 5**: Establish disciplinary steps to be taken when testing shows that a particular CCP is not under regulation.
- » **Principle 6**: Establish testing protocols to ensure that the HACCP device is operating effectively.
- » **Principle 7**: Create documents on all procedures and records that are relevant

These concepts and their implementation.

Good Manufacturing Practices (GMP)

Primary Production

Food establishments shall monitor the pollution of food products / materials from dust, soil, water, feedstuffs, pests, fertilizers, pesticides, veterinary medicines during development, processing, storage and transport, as applicable. Plant and animal welfare are regulated so as not to present a danger to human safety by food intake.

Location and Surroundings

Food establishments shall be situated away from ecologically contaminated areas and manufacturing operations which generate odors, gases, unnecessary soot, dust, smoke, chemical or biological pollution and contaminants which pose a serious risk of contaminating food; flooded areas; areas vulnerable to pest infestation; and areas where waste, either solid or liquid, is present.

Layout and Design of Food Establishment Premises

The architecture of the food establishment shall ensure a forward-looking food preparation / method flow in such a way that cross-contamination from earlier stages of the cycle is eliminated in subsequent phases.

i. Equipment

Equipment and containers which come into contact with food and are used for food handling, transportation, preparing, distribution, packing and serving shall be constructed of materials which do not pose any exposure to the food content. Containers used to store cleaning chemicals and other toxic substances shall be marked and, where necessary, locked to deter unintentional contamination of food.

ii. Facilities

» **Water supply**

Only drinking water that satisfies the criteria of the drinking water standards, with adequate collection, delivery and temperature control facilities, shall be used as an element as well as for food handling, washing, refining and cooking, if applicable. Water holding tanks shall be washed on a regular basis and logs of the same shall be kept.

» **Ice and steam**

Ice and steam used in direct connection with food shall be produced from drinking water and shall comply with the prescribed specifications. Ice and steam shall be made, handled and preserved in such a manner as to prevent them from pollution.

» **Drainage and waste disposal**

The treatment of waste and effluent (solid, liquid and gas) shall be in compliance with the standards of the Environmental Pollution Control Board. Appropriate sanitation, waste management systems and services shall be given. They shall be planned and built in such a manner as to reduce the possibility of contaminating food or drinking water supplies. Waste disposal shall be placed in such a way as not to contaminate the food cycle, storage facilities, the atmosphere within and outside the food establishment. Waste shall be stored in the containers coated and shall not be permitted to accumulate in food handling, food storage and other workplaces.

» **Personnel facilities and toilets**

Service facilities shall provide sufficient means of hygienically washing and drying hands, including washbasins and hot and/or cold water supplies; different laundry facilities of suitable hygienic specification for males and females; and sufficient staff changing facilities. Such facilities shall be conveniently located in such a way that they do not open directly to food processing areas. The rest and refreshment rooms shall be segregated from the food production and distribution areas. Such areas do not contribute directly to food processing, distribution and storage regions.

» **Air quality and ventilation**

Ventilation devices, natural or electronic, like air filters, where appropriate, shall be built and installed such that air does not move from polluted places to healthy areas; reduce airborne pollution of produce; regulation of odours; regulation of atmospheric temperatures and humidity, whenever required, to ensure food health and suitability.

» **Lighting**

Adequate natural or artificial illumination shall be given to enable the undertaking to work in a hygienic manner. Lighting fixtures should be covered, when possible, to insure that food is not tainted by breakages.

Food Operations and Controls

» **Procurement of raw materials**

Any raw material or product shall be approved by the organization whether it is found to include pathogens, harmful micro-organisms, chemicals, veterinary medicines or hazardous, decomposed or extraneous substances which will not be reduced to an suitable amount by usual sorting and/or refining.

» **Storage of raw materials and food**

Food storage facilities shall be built and installed to enable food to be adequately shielded from pollution during preparation; to allow adequate inspection and cleaning; and to prevent pests and harbours.

» **Food Processing, packaging and distribution, temperature control**

The Food Establishment shall establish and sustain a program to insure that time and temperature are efficiently regulated where it is essential to the health and suitability of food. These monitors shall involve the period and temperature of collection, manufacturing, heating, cooling, transportation, packing, delivery and food service to the customer, as appropriate.

» **Precautions against contaminants and cross-contamination**

Systems needs to be implemented to prevent contamination of food materials and food by physical, chemical and microbiological contaminants. Microbiological and chemical examination, effective identification methods for foreign objects shall be used where applicable. Access to food preparation / processing / manufacturing facilities shall be controlled. Furthermore, workers from raw production areas shall not be permitted to step on to the manufacturing areas.

» **Food packaging**

Packaging materials shall have appropriate protection for packaged food items in order to avoid degradation, harm and correct marking. Packaging products or gasses used shall be non-toxic and shall not constitute a hazard to the health and suitability of food under defined storage and usage conditions.

» **Food distribution / service**

Processed, packaged / ready-to-eat food shall be adequately secured during transport and operation. The temperature and humidity required for the preservation of food health and consistency shall be preserved throughout transport and operation. The conveyances / containers shall be built, installed and maintained in such a manner that they can efficiently sustain the appropriate temperature, humidity, environment and other conditions sufficient to protect the food.

Management and supervision

The food establishment shall insure that managers and supervisors have sufficient training, proper awareness and expertise in food hygiene standards and procedures to ensure food health and consistency in their goods, identify food risks, take suitable preventive and corrective measures and ensure that effective inspection and oversight is carried out.

Documentation and records

Appropriate records of food processing / preparation, production / cooking, transportation, delivery, operation, food quality monitoring, hygiene and sanitation, pest control and drug recall shall be established and preserved for a duration of more than one year or more than the shelf-life of the commodity.

Traceability and food products recall

Food company shall insure that appropriate traceability processes are in force from the raw materials to the end product and to the customer in order to cope with any food safety threat and to provide for the full and immediate removal from the market of any food commodity involved.

Good Agricultural Practices (GAP)

Good Agricultural Practices (GAP) are proper food safety principles linked with reducing organic, chemical and physical hazards from the field through the distribution of fresh fruits and vegetables. Such standards come into the following categories: location selection, related land usage, soil, fertilizers (including manure and urban bio-solids), chemicals, staff safety, field and plant ventilation, cooling and transport. Hotlink Annex Great farming activity. Throughout the United States, the recommendations are focused on the current state and regional regulations as well as the Guide to Minimizing Microbial Food Health Hazards throughout Fruits and Vegetables issued by the Food and Drug Administration. Some nations, such as Mexico, have established their own Food Safety Handbook, often focused on the FDA and often sometimes meeting the criteria of the FDA.

Quality Standards

ISO (International Organization for Standardization) is a regional network of national standards bodies (ISO member bodies). The work on the implementation of Universal Requirements shall be carried out by ISO Expert Committees.

Food health is linked to the occurrence of food-borne hazards at the point of ingestion. Since food safety risks may arise at every point of the food chain, proper monitoring of the entire food chain is important. Food safety is therefore ensured through the combined efforts of all the parties involved in the food chain.

The International Standard ISO 22000:2005 defines the criteria for a food health management scheme that incorporates the following widely recognised main elements to maintain food protection through the food chain to the point of final consumption:

The International Standard incorporates the concepts of the Hazard Analysis and Critical Control Point (HACCP) framework and the implementation measures defined by the Codex Alimentarius Commission. This integrates the HACCP program with the Prerequisite Programs (PRPs) by way of auditable specifications. Danger analysis is the secret to an efficient food safety management program, since performing a danger review allows to coordinate the information needed to create an appropriate mix of protection measures. Because ISO 22000 is a standardized food safety management system, it growing be used by any company directly or indirectly participating in the food chain. It refers to all food chain organizations. No matter how complicated a company is or how big it is, ISO 22000 will help insure the health of the food items.

The food chain consists of a number of phases and processes involved in the creation and distribution of food goods. It involves every phase from original output to final usage. More importantly, it involves the production, refining, delivery, storing and handling of both food and nutritional products. The food chain also includes organizations that do not handle food directly. These include organizations that produce animal feed. It also involves companies manufacturing products that would inevitably come into touch with food or product components.

Advantages of ISO 22000: 2005

ISO 22000 will help you achieve the following objectives:

1. Establishment of a Food Safety Management System (FSMS).

2. Ensure all the goods do not have harmful health consequences.

3. Demonstrate accordance with public health standards.

4. Evaluate the food safety requirements of customers.

5. Providing healthy goods and improving consumer loyalty.

6. Import agricultural items and reach overseas markets.

7. Communicate health concerns through the supply chain.

8. Ensure compliance with the food safety policy of the company

Key Elements of ISO 22000

» **Interactive communication**: Communication is important across the food chain to insure that any potential food safety risks are detected and adequately regulated at any phase in the food supply chain. This provides with both upstream and downstream organisations.

» **System management**: The most successful food safety programs need to be developed, implemented within the context of the formal management method and incorporated into the overall management practices of the enterprise.

» **Prerequisite program**: Prerequisite programs are classified into two subcategories. Infrastructure and maintenance initiatives addressing ongoing dimensions of food security. Operational condition systems are structured to reduce the possibility of hazards to the manufacturing or service system.

» **HACCP principles**: The HACCP program is used to monitor the essential control points identified for the removal, avoidance or mitigation of particular food protection hazards from the product as defined through the danger review.

ISO 22000 Food Safety Management System

The FSMS is a series of interrelated steps that define priorities and goals and to accomplish certain goals that are used to coordinate and manage the organisation in relation to food health. Efficient FSMS should be well developed, reported, applied, managed and constantly improved / updated and have its products / services that genuinely fulfill its expected purpose and are healthy and responsive and advanced, research, risk-avoiding and prevention-oriented.

» **Model of the ISO 22000**

The ISO 22000 model is a constant process-based advancement FSMS with a systematic approach to the development, planning, validation, development, implementation, monitoring, verification and improvement of the FSMS.

» **ISO 22000 Certification**

ISO 22000 is meant to be used for certification / registration purposes. In other terms, after an organization has implemented an FSMS that meets the criteria of the ISO, it may submit to a certification agency to inspect the program. Unless, after evaluation, the Certification Agency considers that the program introduced satisfies the criteria of ISO 22000, it will grant an official certificate confirming that the FSMS of the organization meets the specifications of food health.

Nevertheless, qualification is not compulsory. An organization can remain in accordance without being officially approved by an authorized certification entity. The program introduced should be self-assessed and ISO 22000 compatible. However, corporate clients and mutual associates are reluctant to agree that the firm has a successful FSMS because it is not accredited.

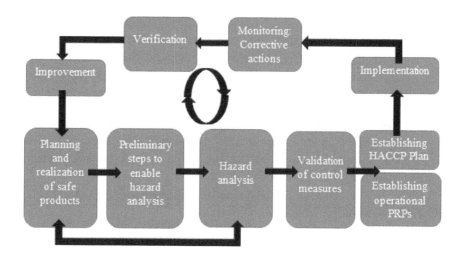

Fig. 17.1: Food Safety Management-a Process model

Source: http://ecoursesonline.iasri.res.in

Benefits of ISO 22000 for users

Organizations that enforce the norm will gain from:

1. Organized and focused contact between business partners;

2. Optimization of capital (both locally and in the food chain);

3. Increased documentation;

4. Great preparation, fewer checking of the post-process;

5. More effective and more active monitoring of food health hazards;

6. Both safety steps subject to a threat analysis;

7. Systematic management of the programs required;

8. Wide application because it focuses on the end result;

9. Strong foundation for decision making;

10. Enhanced due diligence;

11. Control focused on what's needed, and

12. Saving of resources by reducing overlapping system audits.

References

Abdul B. A.; Spence, C. and Hoover, R. (1992). Black polyethylene mulch doubled yield of fresh-market field tomatoes. *Hort Sci.*, **27**(7): 787-789.

Akin-Idowu, P. E.; Ibitoye, D. O. and Ademoyegun, O. T. (2009). Tissue culture as a plant production technique for horticultural crops. *Afri. J. Biotech.*, **8**(16): 3782-3788.

Aldrich, R. A. and Bartok, J. W. (1994). NRAES, Riley, Robb Hall. Green House Engineering. Cornell University, Ithaca, New York.

Al-Kayssi, A. and Mustafa, S. H. (2016). Impact of elevated carbon dioxide on soil heat storage and heat flux under unheated low-tunnels conditions. *J. Envir. Man.*, **182**: 176-186. 10.1016/j.jenvman.2016.07.048.

Ambrosoli, M. (2007). Conservation and Diffusion of Species Diversity in Northern Italy: Peasant Gardeners of the Renaissance and After. In Conan, M. and W. J. Kress, eds. *Bot. Prog., Hort. Inno. Cul. Chan.*, pp: 177-198.

Awadhesh Kumar. (2019). Integration of Vertical Farming and Hydroponics: A Recent Agricultural Trend to Feed the Indian Urban Population in 21st Century. *Acta Sci. Agri.*, **3**(2): 54-59.

Balraj Singh. (2006). Protected cultivation of vegetable crops. Kalyani Publishers, Ludhiana.

Bartok, J. W. (2009). Subirrigation for Greenhouses. http://www.umass.edu/umext/floriculture/fact_sheets/greenhouse_management/jb_subirrigation_whatsnew.pdf

Betelvine Cultivation in India: http://www.agrihortico.com

Bhattacharjee, K. S. and L.C. DE. (2003). Advanced Commercial Floriculture. Aavishkar Publishers, Distributors, Jaipur.

Bhattacharyya, P.; Gupta, S. K.; Banke, S. and Tyagi, B. B. (2010). Crop residue management and development of site specific value addition technology for its use as organic source of plant nutrients. *Organic Farming News Letter,* **6**(3): 6-9.

Bhattacharya, S.; Das S. and Saha, T. (2018). Application of plasticulture in horticulture: A review *The Pharma Inno. J.,* **7**(3): 584-585.

Bose, T. K. and Yadav, L. P. (1989). Commercial Flowers. Naya Prakash, Calcutta.

Brahma Singh. (2014). Advances in Protected Cultivation. New India Publishing Agency. New Delhi.

Brown, S. L. and Brown, J. E. (1992). Effect of plastic mulch color and insecticides on thrips populations and damage to tomato. *Hort Tech.*, **2**(2): 208-211.

Chandra, G. and Sagar, R. L. (2004). Harvesting Green Gold: Cultivation of Betelvine in Sundarban. *Ind. Farmers Digest.*, **37**(3): 5-13.

Cox, D.A Use "BMPs" to Increase Fertilizer Efficiency and Reduce Runoff. http://www.umass.edu/umext/floriculture/fact_sheets/greenhouse_management/bmp.html

Das, S. N.; Pandey, A. K. and Kumar P. (2017). Betelvine cultivation: A new avenue for livelihood security. *HortFlora Res. Spectrum*, **6**(4): 300-303.

Design and maintenance of greenhouse. www.ecourseonline.iasri.res.in

Emmott, C. J. M.; Rohr J. A.; Campoy-Quiles, M., Kirchartz, T.; Urbina, A.; Ekins-Daukes, N. J. and Nelson, J. (2015). *Energy Environ. Sci.*, **8**: 1317.

Farhat Ali and Chitra Srivastava. (2017). Futuristic Urbanism-An overview of Vertical farming and urban agriculture for future cities in India. *Int. J. Adv. Res, Sci., Eng. and Tech.*, **4**(4): 3767-3775.

Floriculturist (Protected Cultivation). (2018). Textbook for Class XI by *National Council of Educational Research and Training.*

Francisco Javier Ferrández-Pastor, Sara Alcañiz-Lucas, Juan Manuel García-Chamizo and Manuel Platero-Horcajadas. (2019). Smart Environments Design on Industrial Automated Greenhouses. *Proceedings*, **31**(36): 1-13.

Good Agricultural Practices for greenhouse vegetable crops: www.fao.org/publications

Gopinath, P.; Vethamoni P. I. and Gomathi, M. (2017). Aeroponics Soilless Cultivation System for Vegetable Crops. *Chem. Sci. Rev. Lett.*, **6**(22): 838-849.

Horticulture: Greenhouse cultivation. www.agritech.tnau.ac.in

IoT Greenhouse: First step towards smart agriculture. Https://R-Stylelab.com/ Company/Blog/Iot/Iot-Agriculture-How-To-Build-Smart-Greenhouse.

Janick, J. (1972). Horticultural science, 2nd edition. W.H. Freeman and company, San Francisco: 564-570.

Jegadeesh, M. and Verapandi, J. (2014). An Innovative Approach on Vertical Farming Techniques. *SSRG-IJAES*, **1**(1): 1-5.

Jindal, Krishnakumar and Madhumeeta. (1994). Diseases of Ornamental Plants in India. Daya Publishing House, New Delhi.

Kheir Al-Kodmany. (2018). The Vertical Farm: A Review of Developments and Implications for the Vertical City. *Buildings*, **8**(24): 1-36.

Kumar, U. and Singh, D. (2013). Protected Horticulture for Sustainable Production. Agriobios Publishers, Jodhpur.

Lakhiar, Imran & Gao, Jianmin & Syed, Tabinda & Chandio, Farman Ali & Buttar, Noman. (2018). Modern plant cultivation technologies in agriculture under controlled environment: A review on aeroponics. *J. Pl. Interact.*, 13. 10.1080/17429145.1472308.

Muthupavithran, S.; Akash, S. and Ranjithkumar, P. (2016). Greenhouse monitering using Internet of Things. *IJIRCSE*, 2(3): 13-19.

Nelson, P. V. (1991). Green House Operation and Management, Bali Publications.

Nirala, S. K.; Suresh R. and Kumar S. (2018). Fertigation Effect on Carnation under Polyhouse in North Bihar Agro-Climatic Conditions. *Int.J.Curr.Microbiol.App. Sci.*, **7**(8): 3746-3755.

Patel, N. L., Chawla, S. L. and Ahlwat, T. R. (2016). Commercial Horticulture. New India Publishing Agency, New Delhi.

Pradeep, T.; Suma, B.; Bhaskar, J. and Satheson, K. N. (2008). Management of Horticultural crops. New India Publishing Agency, New Delhi.

Prasad S. (2005). Greenhouse Management for Horticultural Crops. Agrobios Publishers, Jodhpur.

Prasad, S.; Singh, D. and Bhardwaj, R. L. (2012). A book on Hi-Tech Horticulture. Agrobios Publishers, Jodhpur

Rashmi, M. R. and Pavithra, M. P. (2018). Vertical farming: A concept. *Int. J. Engi. Tech.*, **4**(3): 500-506.

Reddy P. Parvatha. (2003). Protected Cultivation. Springer Publications. USA.

Reddy, P. Parvatha. (2011). Sustainable crop protection under Protected Cultivation. Springer Publications. USA.

Reddy, Y. T. and Reddy S. G. (2016). A book on Principles of Agronomy. Kalyani Publishers, New Delhi.

Sathyanarayana Reddy, B.; Janakiram, T.; Balaji, S. Kulkarni and Misra, R. L. (2004). Hi-Tech Floriculture. Indian Society of Ornamental Horticulture, IARI, New Delhi.

Scott, S. J.; Mc Leod, P..; Montgomery, F. W. and Hander, C. A. (1989). Influence of reflective mulch on incidence of thrips (Thysanoptera: Thripidae: Phlaeothripidae) in staked tomatoes. *J. Entomo. Sci.*, **24**(4): 422-427.

Singh, A. K. (2006). Flower Crops Cultivation and Management. New India Publishing Agency, New Delhi.

Singh, A. K. and Das, D. (2018). Integrated vertical farming system. *Ind. Farm.*, **68**(6): 23–24.

Singh, D. K. (2011). A book on Hi-Tech Horticulture. Agrotech Publishing Academy, Udaipur.

Singh, R.R., L.K. Meena and Paramveer Singh. 2017. High Tech Nursery Management in Horticultural Crops: A Way for Enhancing Income. *Int.J.Curr.Microbiol. App.Sci.*, **6**(6): 3162-3172.

Sonawane, M. S. (2018). Status of Vertical Farming in India. *Int. Arch. App. Sci. Tech.*, **9**(4): 122-125.

Sudheer, K. P.; Sureshkumar, P. K.; Geethu, Krishnan and Sheeja, P. S. 2018. Lecture Notes and Lab Manual of Protected Cultivation and Post Harvest Technology.

Syamal, M. M. (2014). Commercial Floriculture. Jaya Publishing House, Delhi.

Tapas Bhattacharyya; Haldankar, P. M.; Patil, V. K.; Salvi, B. R.; Haldavanekar, P. C.; Pujari K. H. and Dosani A. A. (2015). Hi-tech Horticulture: Pros and Cons. *Ind. J. Ferti.*, **13**: 46-58.

Vertical Farming: www.attra.ncat.org

Wien, H. C. and Minotti, P. L. (1987). Growth, yield, and nutrient uptake of transplanted fresh-market tomatoes as affected by plastic mulch and initial nitrogen rate. *Journal of the Ame. Soc. for Hort. Sci.*, **112**(5):759-763.

Yadav, I. S. and Choudary, M. L. (1997). Progressive Floriculture-Production Technologies of Important Commercial Flower Crops. The House of Sarpan, Bengaluru.